Birds *of* Kentucky

Field Guide

Stan Tekiela

Adventure Publications
Cambridge, Minnesota

Edited by Sandy Livoti and Dan Downing

Cover, book design and illustrations by Jonathan Norberg

Range maps produced by Anthony Hertzel

Cover photo: Northern Cardinal by **Stan Tekiela**
All photos by Stan Tekiela except p. 284 (female & male) by **Agami Photo Agency/Shutterstock.com**; p. 276 (juvenile) by **Rick & Nora Bowers**; pp. 40 (soaring), 200 (female), 286 (female) by **Kevin T. Karlson**; pp. 80 (female), 122 (female) by **Brian E. Kushner/Shutterstock.com**; pp. 270 (female), 296 (female) by **Maslowski Wildlife Productions**; p. 160 by **Paul Reeves Photography/Shutterstock.com**; p. 40 (drying) by **Jay Pierstorff/Shutterstock.com**; p. 228 (displaying) by **Hartmut Walter**; pp. 40 (juvenile), 42 (juvenile), 144 (juvenile & in-flight juvenile), 236 (juvenile), 238 (in-flight juvenile) by **Brian K. Wheeler**; and pp. 188 (female), 234 (main), 240 (female), 292 (female) by **Jim Zipp**.

To the best of the publisher's knowledge, all photos were of live birds. Some were photographed in a controlled condition.

10 9 8 7 6 5 4 3 2 1
Birds of Kentucky Field Guide
First Edition 2001
Second Edition 2022
Copyright © 2001 and 2022 by Stan Tekiela
Published by Adventure Publications
An imprint of AdventureKEEN
310 Garfield Street South
Cambridge, Minnesota 55008
(800) 678-7006
www.adventurepublications.net
Printed in China
ISBN 978-1-64755-297-8 (pbk.); ISBN 978-1-64755-298-5 (ebook)

TABLE OF CONTENTS

WHAT'S NEW?

It is hard to believe that it's been more than 20 years since the debut of *Birds of Kentucky Field Guide*. This critically acclaimed field guide has helped countless people identify and enjoy the birds that we love. Now, in this expanded second edition, *Birds of Kentucky Field Guide* has many new and exciting changes and a fresh look, while retaining the same familiar, easy-to-use format.

To help you identify even more birds in Kentucky, I have added 7 new species and more than 150 new color photographs. All of the range maps have been meticulously reviewed, and many updates have been made to reflect the ever-changing movements of the birds.

Everyone's favorite section, "Stan's Notes," has been expanded to include even more natural history information. "Compare" sections have been updated to help ensure that you correctly identify your bird, and additional feeder information has been added to help with bird feeding. I hope you will enjoy this great new edition as you continue to learn about and appreciate our Kentucky birds!

WHY WATCH BIRDS IN KENTUCKY?

Millions of people have discovered bird feeding. Its a simple and enjoyable way to bring the beauty of birds closer to your home. Watching birds at your feeder often leads to a lifetime pursuit of bird identification. The *Birds of Kentucky Field Guide* is for those who want to identify common birds of Kentucky.

There are over 1,100 species of birds found in North America. In Kentucky alone there have been more than 360 different kinds of birds recorded through the years. These bird sightings were diligently recorded by hundreds of bird watchers and became part of the official state record. From these valuable records, I have chosen 118 of the most common birds of Kentucky to include in this field guide.

Bird watching, often called birding, is one of the most popular activities in America. Its outstanding appeal in Kentucky is due, in part, to an unusually rich and abundant birdlife. Why are there so many birds? One reason is open space. Kentucky is more than 40,000 square miles (104,000 sq. km), making it the thirty-seventh-largest state. Despite its size, only about 4.5 million people call Kentucky home. On average, thats only 111 people per square mile (43 per sq. km). And many of these people are located in and around three major metropolitan areas.

Open space is not the only reason there is such an abundance of birds. Its also the diversity of habitat. The state can be broken into several distinct habitats, each of which supports a different group of birds. Water also plays a big part of Kentucky's bird populations. The Ohio River comprises the entire northern border of the state. The Big Sandy and Tug Fork Rivers form much of the eastern border, and the Mississippi River forms the western boundary. All of these riverways provide critical habitats for a wide variety of birds. For example, the floodplain forest along the Mississippi River in western Kentucky is a good place to see wetland birds such as the Belted Kingfisher.

The south-central portion of the state (Highland Rim), which was originally a prairie habitat, is now agricultural with many open-country birds such as the Horned Lark and Eastern Kingbird.

Eastern Kentucky (Cumberland Plateau and Appalachian Plateau) is known for its extensive tracts of forest, rolling hills and mountain ranges. The forested valleys and ridges are home to many birds such as the Scarlet Tanager.

Varying habitats in Kentucky also mean variations in weather. Since the state extends over 400 miles (644 km) from east to west, the weather ranges greatly. From the high and relatively snowy Appalachian Plateau in the east to the steamy summers in lowland western Kentucky, there are birds to watch in every season. Whether witnessing a migration of hawks in the fall or welcoming back the hummingbirds in the spring, there is variety and excitement in birding as each season turns to the next.

OBSERVE WITH A STRATEGY: TIPS FOR IDENTIFYING BIRDS

Identifying birds isn't as difficult as you might think. By simply following a few basic strategies, you can increase your chances of successfully identifying most birds that you see. One of the first and easiest things to do when you see a new bird is to note **its color**. This field guide is organized by color, so simply turn to the right color section to find it.

Next, note the **size of the bird.** A strategy to quickly estimate size is to compare different birds. Pick a small, a medium and a large bird. Select an American Robin as the medium bird. Measured from bill tip to tail tip, a robin is 10 inches (25 cm). Now select two other birds, one smaller and one larger. Good choices are a House Sparrow, at about 6 inches (15 cm), and an American Crow, around 18 inches (45 cm). When you see a species you don't know, you can now quickly ask yourself, "Is it larger than a sparrow but smaller than a robin?" When you look in your field guide to identify your bird, you would check the species that are roughly 6–10 inches (15–25 cm). This will help to narrow your choices.

Next, note the **size, shape and color of the bill.** Is it long or short, thick or thin, pointed or blunt, curved or straight? Seed-eating birds, such as Northern Cardinals, have bills that are thick and strong enough to crack even the toughest seeds. Birds that sip nectar, such as Ruby-throated Hummingbirds, need long, thin bills to reach deep into flowers. Hawks and owls tear their prey with very sharp, curving bills. Sometimes, just noting the bill shape can help you decide whether the bird is a woodpecker, finch, grosbeak, blackbird or bird of prey.

Next, take a look around and note the **habitat** in which you see the bird. Is it wading in a marsh? Walking along a riverbank? Soaring in the sky? Is it perched high in the trees or hopping along the forest floor? Because of diet and habitat preferences, you'll often see robins hopping on the ground but not usually eating seeds at a feeder. Or you'll see a Rose-breasted Grosbeak on a tree branch but not climbing headfirst down the trunk, like a White-breasted Nuthatch would.

Noticing **what the bird is eating** will give you another clue to help you identify the species. Feeding is a big part of any bird's life. Fully one-third of all bird activity revolves around searching for food, catching prey and eating. While birds don't always follow all the rules of their diet, you can make some general assumptions. Northern Flickers, for instance, feed on ants and other insects, so you wouldn't expect to see them visiting a seed feeder. Other birds, such as Barn and Tree Swallows, eat flying insects and spend hours swooping and diving to catch a meal.

Sometimes you can identify a bird by **the way it perches.** Body posture can help you differentiate between an American Crow and a Red-tailed Hawk, for example. Crows lean forward over their feet on a branch, while hawks perch in a vertical position. Consider posture the next time you see an unidentified large bird in a tree.

Birds in flight are harder to identify, but noting the **wing size and shape** will help. Wing size is in direct proportion to body size, weight and type of flight. Wing shape determines whether the bird flies fast and with precision, or slowly and less precisely. Barn Swallows, for instance, have short, pointed wings that slice through the air, enabling swift, accurate flight. Turkey Vultures have long, broad wings for soaring on warm updrafts. House Finches have short, rounded wings, helping them to flit through thick tangles of branches.

Some bird species have a unique **pattern of flight** that can help in identification. American Goldfinches fly in a distinctive undulating pattern that makes it look like they're riding a roller coaster.

While it's not easy to make all of these observations in the short time you often have to watch a "mystery" bird, practicing these identification methods will greatly expand your birding skills. To further improve your skills, seek the guidance of a more experienced birder who can answer your questions on the spot.

BIRD BASICS

It's easier to identify birds and communicate about them if you know the names of the different parts of a bird. For instance, it's more effective to use the word "crest" to indicate the set of extra-long feathers on top of a Northern Cardinal's head than to try to describe it.

The following illustration points out the basic parts of a bird. Because it is a composite of many birds, it shouldn't be confused with any actual bird.

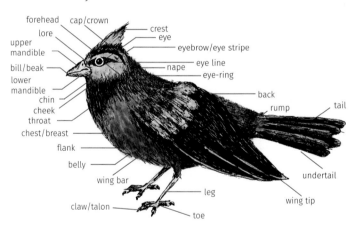

Bird Color Variables

No other animal has a color palette like a bird's. Brilliant blues, lemon yellows, showy reds and iridescent greens are common in the bird world. In general, male birds are more colorful than their female counterparts. This helps males attract a mate, essentially saying, "Hey, look at me!" Color calls attention to a male's health as well. The better the condition of his feathers, the better his food source, territory and potential for mating.

Male and female birds that don't look like each other are called sexually dimorphic, meaning "two forms." Dimorphic females often have a nondescript dull color, as seen in Indigo Buntings. Muted tones help females hide during the weeks of motionless incubation and draw less attention to them when they're out feeding or taking a break from the rigors of raising the young.

The males of some species, such as the Downy Woodpecker, Blue Jay and Bald Eagle, look nearly identical to the females. In woodpeckers, the sexes are differentiated by only a red mark, or sometimes a yellow mark. Depending on the species, the mark may be on top of the head, on the face or nape of neck, or just behind the bill.

During the first year, juvenile birds often look like their mothers. Since brightly colored feathers are used mainly for attracting a mate, young non-breeding males don't have a need for colorful plumage. It's not until the first spring molt (or several years later, depending on the species) that young males obtain their breeding colors.

Both breeding and winter plumages are the result of molting. Molting is the process of dropping old, worn feathers and replacing them with new ones. All birds molt, typically twice a year, with the spring molt usually occurring in late winter. At this time, most birds produce their brighter breeding plumage, which lasts throughout the summer.

Winter plumage is the result of the late summer molt, which serves a couple of important functions. First, it adds feathers for warmth in the coming winter season. Second, in some species it produces feathers that tend to be drab in color, which helps to camouflage the birds and hide them from predators. The winter plumage of the male American Goldfinch, for example, is olive-brown, unlike its canary-yellow breeding color during summer. Luckily for us, some birds, such as the male Northern Cardinal, retain their bright summer colors all year long.

Bird Nests

Bird nests are a true feat of engineering. Imagine constructing a home that's strong enough to weather storms, large enough to hold your entire family, insulated enough to shelter them from cold and heat, and waterproof enough to keep out rain. Think about building it without blueprints or directions and using mainly your feet. Birds do this!

Before building, birds must select an appropriate site. In some species, such as the House Wren, the male picks out several potential sites and assembles small twigs in each. The "extra" nests, called dummy nests, discourage other birds from using any nearby cavities for their nests. The male takes the female around and shows her the choices. After choosing her favorite, she finishes the construction.

In other species, such as the Baltimore Oriole, the female selects the site and builds the nest, while the male offers an occasional suggestion. Each bird species has its own nest-building routine that is strictly followed.

As you can see in these illustrations, birds build a wide variety of nest types.

| ground nest | platform nest | cup nest | pendulous nest | cavity nest |

Nesting material often consists of natural items found in the immediate area. Most nests consist of plant fibers (such as bark from grapevines), sticks, mud, dried grass, feathers, fur, or soft, fuzzy tufts from thistle. Some birds, including Ruby-throated Hummingbirds, use spiderwebs to glue nest materials together.

Transportation of nesting material is limited to the amount a bird can hold or carry. Birds must make many trips afield to gather enough material to complete a nest. Most nests take four days or more, and hundreds, if not thousands, of trips to build.

A **ground nest** can be a mound of vegetation on the ground or in water. It can also be just a simple, shallow depression scraped out in earth, stones or sand. Killdeer and Horned Larks scrape out ground nests without adding any nesting material.

The **platform nest** represents a much more complex type of construction. Typically built with twigs or sticks and branches, this nest forms a platform and has a depression in the center to nestle the eggs. Platform nests can be in trees; on balconies, cliffs, bridges, or man-made platforms; and even in flowerpots. They often provide space for the adventurous young and function as a landing platform for the parents.

Mourning Doves and herons don't anchor their platform nests to trees, so these can tumble from branches during high winds and storms. Hawks, eagles, Ospreys and other birds construct sturdier platform nests with large sticks and branches.

Other platform nests are constructed on the ground with mud, grass and other vegetation from the area. Many waterfowl build platform nests on the ground near or in water. A **floating platform nest** moves with the water level, preventing the nest, eggs and birds from being flooded.

Three-quarters of all songbirds construct a **cup nest,** which is a modified platform nest. The supporting platform is built first and attached firmly to a tree, shrub, or rock ledge or the ground. Next, the sides are constructed with grass, small twigs, bark or leaves, which are woven together and often glued with mud for added strength. The inner cup can be lined with down feathers, animal fur or hair, or soft plant materials and is contoured last.

The **pendulous nest** is an unusual nest that looks like a sock hanging from a branch. Attached to the end of small branches of trees, this unique nest is inaccessible to most predators and often waves wildly in a breeze.

Woven tightly with plant fibers, the pendulous nest is strong and watertight and takes up to a week to build. A small opening at the top or on the side allows parents access to the grass-lined interior. More commonly used by tropical birds, this complex nest has also been mastered by orioles and kinglets. It must be one heck of a ride to be inside one of these nests during a windy spring thunderstorm!

The **cavity nest** is used by many species of birds, most notably woodpeckers and Eastern Bluebirds. A cavity nest is often excavated from a branch or tree trunk and offers shelter from storms, sun, cold and predators. A small entrance hole in a tree can lead to a nest chamber that is up to a safe 10 inches (25 cm) deep.

Typically made by woodpeckers, cavity nests are usually used only once by the builder. Nest cavities can be used for many subsequent years by such inhabitants as mergansers, Tree Swallows and bluebirds. Kingfishers, on the other hand, can dig a tunnel up to 4 feet (1 m) long in a riverbank. The nest chamber at the end of the tunnel is already well insulated, so it's usually only sparsely lined.

One of the most clever of all nests is the **no nest,** or daycare nest. Parasitic birds, such as Brown-headed Cowbirds, don't build their own nests. Instead, the egg-laden female searches out the nest of another bird and sneaks in to lay an egg while the host mother isn't looking.

A mother cowbird wastes no energy building a nest only to have it raided by a predator. Laying her eggs in the nests of other birds transfers the responsibility of raising her young to the host. When she lays her eggs in several nests, the chances increase that at least one of her babies will live to maturity.

Who Builds the Nest?

Generally, the female bird constructs the nest. She gathers the materials and does the building, with an occasional visit from her mate to check on progress. In some species, both parents contribute equally to nest building. The male may forage for sticks, grass or mud, but it is the female that often fashions the nest. Only rarely does a male build a nest by himself.

Fledging

Fledging is the time between hatching and flight, or leaving the nest. Some species of birds are **precocial,** meaning they leave the nest within hours of hatching, though it may be weeks before they can fly. This is common in waterfowl and shorebirds.

Baby birds that hatch naked and blind need to stay in the nest for a few weeks (these birds are **altricial**). Baby birds that are still in the nest are **nestlings.** Until birds start to fly, they are called **fledglings.**

Why Birds Migrate

Why do so many species of birds migrate? The short answer is simple: food. Birds migrate to locations with abundant food, as it is easier to breed where there is food than where food is scarce. Summer Tanagers, for instance, are **complete migrators** that fly from the tropics of South America to nest in the forests of North America, where billions of newly hatched insects are available to feed to their young.

Other migrators, such as some birds of prey, migrate back to northern regions in spring. In these locations, they hunt mice, voles and other small rodents that are beginning to breed.

Complete migrators have a set time and pattern of migration. Every year at nearly the same time, they head to a specific wintering ground. Complete migrators may travel great distances, sometimes 15,000 miles (24,100 km) or more in one year.

Complete migration doesn't necessarily mean flying from Kentucky to a tropical destination. Dark-eyed Juncos, for example, are complete migrators that move from the far reaches of Canada to spend the winter here in Kentucky. This trip is still considered complete migration.

Complete migrators have many interesting aspects. In spring, males often leave a few weeks before the females, arriving early to scope out possibilities for nesting sites and food sources, and to begin to defend territories. The females arrive several weeks later. In many species, the females and their young leave earlier in the fall, often up to four weeks before the adult males.

Other species, such as the American Goldfinch, are **partial migrators.** These birds usually wait until their food supplies dwindle before flying south. Unlike complete migrators, partial migrators move only far enough south, or sometimes east and west, to find abundant food. In some years it might be only a few hundred miles, while in other years it can be as much as a thousand. This kind of migration, dependent on weather and the availability of food, is sometimes called seasonal movement.

Unlike the predictable complete migrators or partial migrators, **irruptive migrators** can move every third to fifth year or, in some cases, in consecutive years. These migrations are triggered when times are tough and food is scarce. Red-breasted Nuthatches are irruptive migrators. They leave their normal northern range in search of more food or in response to overpopulation.

Many other birds don't migrate at all. Carolina Chickadees, for example, are **non-migrators** that remain in their habitat all year long and just move around as necessary to find food.

How Do Birds Migrate?

One of the many secrets of migration is fat. While most people are fighting the ongoing battle of the bulge, birds intentionally gorge themselves to gain as much fat as possible without losing the ability to fly. Fat provides the greatest amount of energy per unit of weight. In the same way that your car needs gas, birds are propelled by fat and stall without it.

During long migratory flights, fat deposits are used up quickly, and birds need to stop to refuel. This is when backyard bird feeding stations and undeveloped, natural spaces around our towns and cities are especially important. Some birds require up to 2–3 days of constant feeding to build their fat reserves before continuing their seasonal trip.

Many birds, such as most eagles, hawks, Ospreys, falcons and vultures, migrate during the day. Larger birds can hold more body fat, go longer without eating and take longer to migrate. These birds glide along on rising columns of warm air, called thermals, that hold them aloft while they slowly make their way north or south. They generally rest at night and hunt early in the morning before the sun has a chance to warm the land and create good soaring conditions. Daytime migrators use a combination of landforms, rivers, and the rising and setting sun to guide them in the right direction.

The majority of small birds, called **passerines,** migrate at night. Studies show that some use the stars to navigate. Others use the setting sun, and still others, such as pigeons, use Earth's magnetic field to guide them north or south.

While flying at night may not seem like a good idea, it's actually safer. First, there are fewer avian predators hunting for birds at night. Second, night travel allows time during the day to find food in unfamiliar surroundings. Third, wind patterns at night tend to be flat, or laminar. Flat winds don't have the turbulence of daytime winds and can help push the smaller birds along.

HOW TO USE THIS GUIDE

To help you quickly and easily identify birds, this field guide is organized by color. Refer to the color key on the first page, note the color of the bird, and turn to that section. For example, the male Red-headed Woodpecker is black-and-white with a red head. Because the bird is mostly black-and-white, it will be found in the black-and-white section.

Each color section is also arranged by size, generally with the smaller birds first. Sections may also incorporate the average size in a range, which in some cases reflects size differences between male and female birds. Flip through the pages in the color section to find the bird. If you already know the name of the bird, check the index for the page number.

In some species, the male and female are very different in color. In others, the breeding and winter plumage colors differ. These species will have an inset photograph with a page reference and will be found in two color sections.

You will find a variety of information in the bird description sections. To learn more, turn to the sample on pp. 22–23.

Range Maps

Range maps are included for each bird. Colored areas indicate where the bird is frequently found. The colors represent the presence of a species during a specific season, not the density, or amount, of birds in the area. Green is used for summer, blue for winter, red for year-round and yellow for migration.

While every effort has been made to depict accurate ranges, these are constantly in flux due to a variety of factors. Changing weather, habitat, species abundance and availability of vital resources, such as food and water, can affect the migration and movement of local populations, causing birds to be found in areas that are atypical for the species. So please use the maps as intended—as general guides only.

male

female
p. 107

Common Name

YEAR-ROUND
SUMMER
MIGRATION
WINTER

Size: measurement is from head to tip of tail; wingspan may be listed as well

Male: brief description of the male bird; may include breeding, winter or other plumages

Female: brief description of the female bird, which is sometimes different from the male

Juvenile: brief description of the juvenile bird, which often looks like the adult female

Nest: kind of nest the bird builds to raise its young; who builds it; number of broods per year

Eggs: number of eggs you might expect to see in a nest; color and marking

Incubation: average days the parents spend incubating the eggs; who does the incubation

Fledging: average days the young spend in the nest after hatching but before they leave the nest; who does the most "childcare" and feeding

Migration: type of migrator: complete (seasonal, consistent), partial (seasonal, destination varies), irruptive (unpredictable, depends on the food supply) or non-migrator

Food: what the bird eats most of the time (e.g., seeds, insects, fruit, nectar, small mammals, fish) and whether it typically comes to a bird feeder

Compare: notes about other birds that look similar and the pages on which they can be found; may include extra information to aid in identification

Stan's Notes: Interesting natural history information. This could be something to look or listen for or something to help positively identify the bird. Also includes remarkable features.

female
p. 129

male

YEAR-ROUND

Eastern Towhee
Pipilo erythrophthalmus

Size: 7–8" (18–20 cm)

Male: Mostly black with rusty-brown sides and a white belly. Long black tail with a white tip. Short, stout, pointed bill and rich, red eyes. White wing patches flash in flight.

Female: similar to male but brown instead of black

Juvenile: light brown, a heavily streaked head, chest and belly, long dark tail with white tip

Nest: cup; female builds; 2 broods per year

Eggs: 3–4; creamy white with brown markings

Incubation: 12–13 days; female incubates

Fledging: 10–12 days; male and female feed the young

Migration: non-migrator to partial; moves around to find food

Food: insects, seeds, fruit; visits ground feeders

Compare: American Robin (p. 227) is slightly larger. The Gray Catbird (p. 225) lacks a black "hood" and rusty sides. Male Rose-breasted Grosbeak (p. 51) has a rosy patch in center of chest. Male Orchard Oriole (p. 265) is the same size, but unlike ground-dwelling Towhee, is often high in trees.

Stan's Notes: Named for its distinctive "tow-hee" call (given by both sexes) but known mostly for its other characteristic call, which sounds like "drink-your-tea!" Will hop backward with both feet (bilateral scratching), raking up leaf litter to locate insects and seeds. The female broods, but male does the most feeding of young. In southern states, some have red eyes and others have white eyes.

female
p. 125

male

Brown-headed Cowbird
Molothrus ater

YEAR-ROUND

Size: 7½" (19 cm)

Male: Glossy black with a chocolate-brown head. Dark eyes. Pointed, sharp gray bill.

Female: dull brown with a pointed, sharp, gray bill

Juvenile: similar to female but with dull-gray plumage and a streaked chest

Nest: no nest; lays eggs in nests of other birds

Eggs: 5–7; white with brown markings

Incubation: 10–13 days; host bird incubates eggs

Fledging: 10–11 days; host birds feed the young

Migration: non-migrator; moves around to find food

Food: insects, seeds; will come to seed feeders

Compare: The male Red-winged Blackbird (p. 31) is slightly larger with red-and-yellow patches on upper wings. Common Grackle (p. 33) has a long tail and lacks the brown head. European Starling (p. 29) has a shorter tail.

Stan's Notes: Cowbirds are members of the blackbird family. Known as brood parasites, Brown-headed Cowbirds are the only parasitic birds in Kentucky. Brood parasites lay their eggs in the nests of other birds, leaving the host birds to raise their young. Cowbirds are known to have laid their eggs in the nests of over 200 species of birds. While some birds reject cowbird eggs, most incubate them and raise the young, even to the exclusion of their own. Look for warblers and other birds feeding young birds twice their own size. Named "Cowbird" for its habit of following bison and cattle herds to feed on insects flushed up by the animals.

winter

breeding

European Starling
Sturnus vulgaris

YEAR-ROUND

Size: 7½" (19 cm)

Male: Glittering, iridescent purplish black in spring and summer; duller and speckled with white in fall and winter. Long, pointed, yellow bill in spring; gray in fall. Pointed wings. Short tail.

Female: same as male

Juvenile: similar to adults, with grayish-brown plumage and a streaked chest

Nest: cavity; male and female line cavity; 2 broods per year

Eggs: 4–6; bluish with brown markings

Incubation: 12–14 days; female and male incubate

Fledging: 18–20 days; female and male feed the young

Migration: non-migrator to partial migrator; moves around to find food

Food: insects, seeds, fruit; visits seed or suet feeders

Compare: The Common Grackle (p. 33) has a long tail. Male Brown-headed Cowbird (p. 27) has a brown head. Look for the shiny, dark feathers to help identify the European Starling.

Stan's Notes: One of our most numerous songbirds. Mimics the songs of up to 20 bird species and imitates sounds, including the human voice. Jaws are more powerful when opening than when closing, enabling the bird to pry open crevices to find insects. Often displaces woodpeckers, chickadees and other cavity-nesting birds. Large families gather with blackbirds in the fall. Not a native bird; 100 starlings were introduced to New York City in 1890–91 from Europe. Bill changes color in spring and fall.

female
p. 141

male

Red-winged Blackbird
Agelaius phoeniceus

YEAR-ROUND

Size: 8½" (22 cm)

Male: Jet black with red-and-yellow patches (epaulets) on upper wings. Pointed black bill.

Female: heavily streaked brown with a pointed brown bill and white eyebrows

Juvenile: same as female

Nest: cup; female builds; 2–3 broods per year

Eggs: 3–4; bluish green with brown markings

Incubation: 10–12 days; female incubates

Fledging: 11–14 days; female and male feed the young

Migration: non-migrator to partial; moves around to find food

Food: seeds, insects; visits seed and suet feeders

Compare: The male Brown-headed Cowbird (p. 27) is smaller and glossier and has a brown head. Bold red-and-yellow epaulets distinguish the male Red-winged from other blackbirds.

Stan's Notes: One of the most widespread and numerous birds in Kentucky. In fall and winter, migrant and resident Red-wings gather in huge numbers (thousands) with other blackbirds to feed in agricultural fields, marshes, wetlands, and lakes and rivers. Flocks with as many as 10,000 birds have been reported. Males defend their territory by singing from the tops of surrounding vegetation. The male repeats his call from the top of a cattail while showing off his red-and-yellow shoulder patches. The female chooses a mate and often builds her nest over shallow water in a thick stand of cattails. The male can be aggressive when defending the nest. Feeds mostly on seeds in spring and fall, and insects throughout the summer.

Common Grackle
Quiscalus quiscula

Size: 11–13" (28–33 cm)

Male: Large, iridescent blackbird with bluish-black head and purplish-brown body. Long black tail. Long, thin bill and bright-golden eyes.

Female: similar to male but smaller and duller

Juvenile: similar to female

Nest: cup; female builds; 2 broods per year

Eggs: 4–5; greenish white with brown markings

Incubation: 13–14 days; female incubates

Fledging: 16–20 days; female and male feed the young

Migration: non-migrator to partial; moves around to find food

Food: fruit, seeds, insects; will come to seed and suet feeders

Compare: The European Starling (p. 29) is much smaller with a speckled appearance, and a yellow bill during breeding season. The male Red-winged Blackbird (p. 31) has red-and-yellow wing markings (epaulets).

Stan's Notes: Usually nests in small colonies of up to 75 pairs but travels with other blackbird species in large flocks. Known to feed in farm fields. The common name is derived from the Latin word *gracula*, meaning "jackdaw," another species of bird and a term that can refer to any bird in the *Quiscalus* genus. The male holds his tail in a deep V shape during flight. The flight pattern is usually level, as opposed to an undulating movement. Unlike most birds, it has larger muscles for opening its mouth than for closing it, enabling it to pry crevices apart to find hidden insects.

Common Gallinule
Gallinula galeata

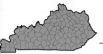

SUMMER
MIGRATION

Size: 13–15" (33–38 cm)

Male: Nearly black overall with yellow-tipped red bill. Red forehead. Thin line of white along sides. Yellowish-green legs.

Female: same as male

Juvenile: same as adult, but brown with white throat and dirty-yellow legs

Nest: ground; female and male build; 1–2 broods per year

Eggs: 2–10; brown with dark markings

Incubation: 19–22 days; female and male incubate

Fledging: 40–50 days; female and male feed the young

Migration: complete, to southern coastal states

Food: insects, snails, seeds, green leaves, fruit

Compare: American Coot (p. 37) is similar in size but lacks the distinctive yellow-tipped bill and red forehead of Common Gallinule.

Stan's Notes: Also known as Mud Hen or Pond Chicken. A nearly all-black duck-like bird often seen in freshwater marshes and lakes. Walks on floating vegetation or swims while hunting for insects. Females known to lay eggs in other gallinule nests in addition to their own. Builds its nest with cattails and bulrushes and sometimes takes an old nest in a low shrub. A cooperative breeder, having young of first brood help raise young of second. Young leave nest usually within a few hours after hatching but stay with the family for a couple months. Young ride on backs of adults.

American Coot
Fulica americana

MIGRATION WINTER

Size: 13–16" (33–40 cm)

Male: Gray-to-black waterbird. Duck-like white bill with a dark band near the tip and a small red patch near the eyes. Small white patch near base of tail. Green legs and feet. Red eyes.

Female: same as male

Juvenile: much paler than adults, with a gray bill

Nest: floating platform; female and male construct; 1 brood per year

Eggs: 9–12; pinkish buff with brown markings

Incubation: 21–25 days; female and male incubate

Fledging: 49–52 days; female and male feed young

Migration: complete, to southern states, Mexico and Central America

Food: insects, aquatic plants

Compare: Smaller than most waterfowl, it is the only black, duck-like bird with a white bill.

Stan's Notes: Usually seen in large flocks on open water. Not a duck, as it has large lobed toes instead of webbed feet. An excellent diver and swimmer, bobbing its head as it swims. A favorite food of Bald Eagles. It is not often seen in flight, unless it's trying to escape from an eagle. To take off, it scrambles across the surface of the water, flapping its wings. Gives a unique series of creaks, groans and clicks. Anchors its floating platform nest to vegetation. Huge flocks with as many as 1,000 birds gather for migration. Migrates at night. The common name "Coot" comes from the Middle English word *coote*, which was used to describe various waterfowl. Also called Mud Hen.

in flight

American Crow
Corvus brachyrhynchos

Size: 18" (45 cm)

Male: All-black bird with black bill, legs and feet. Can have a purple sheen in direct sunlight.

Female: same as male

Juvenile: same as adult

Nest: platform; female builds; 1 brood per year

Eggs: 4–6; bluish to olive-green with brown marks

Incubation: 18 days; female incubates

Fledging: 28–35 days; female and male feed the young

Migration: non-migrator to partial migrator; moves around to find food

Food: fruit, insects, mammals, fish, carrion; will come to seed and suet feeders

Compare: Black Vulture (p. 41) and Turkey Vulture (p. 43) are much larger, and Turkey Vulture has a naked red head.

Stan's Notes: One of the most recognizable birds in Kentucky, found in all habitats. Imitates other birds and human voices. One of the smartest of all birds and very social, often entertaining itself by provoking chases with other birds. Eats roadkill but is rarely hit by vehicles. Can live as long as 20 years. Often reuses its nest every year if it's not taken over by a Great Horned Owl. Unmated birds, known as helpers, help to raise the young. Extended families roost together at night, dispersing daily to hunt. Cannot soar on thermals; flaps constantly and glides downward. Gathers in huge communal flocks of up to 10,000 birds in winter.

drying

juvenile

soaring

YEAR-ROUND

Black Vulture
Coragyps atratus

Size: 25–28" (63–71 cm); up to 5¼' wingspan

Male: Black with dark-gray head and legs. Short tail. In flight, all black with light-gray wing tips.

Female: same as male

Juvenile: similar to adult

Nest: no nest, on a stump or on ground, or takes an abandoned nest; 1 brood per year

Eggs: 1–3; light green with dark markings

Incubation: 37–45 days; female and male incubate

Fledging: 75–80 days; female and male feed the young

Migration: non-migrator; moves around to find food

Food: carrion; occasionally will capture small live mammals

Compare: Turkey Vulture (p. 43) is slightly larger, with a bright-red head. Turkey Vulture has two-toned wings with a black leading edge and light-gray trailing edge. The Black Vulture has shorter, gray-tipped wings and a shorter tail.

Stan's Notes: Also called Black Buzzard. A more gregarious bird than the Turkey Vulture. In flight, the Black Vulture holds its wings straight out to its sides unlike the Turkey Vulture, which holds its wings in a V pattern. More aggressive while feeding but less skilled at finding carrion than the Turkey Vulture, it is thought the Black Vulture's sense of smell is less developed. Families stay together up to a year. Often nests and roosts with other Black Vultures. If startled, especially at the nest, it regurgitates with power and accuracy. Like other vultures, it often spreads its wings to warm up in the morning or dry out after a rain (see inset).

soaring

juvenile

drying

Turkey Vulture

Cathartes aura

Size: 26–32" (66–80 cm); up to 6' wingspan

Male: Large and black with a naked red head and legs. In flight, wings are two-toned with a black leading edge and a gray trailing edge. Wing tips end in finger-like projections. Tail is long and squared. Ivory bill.

Female: same as male but slightly smaller

Juvenile: similar to adults, with a gray-to-blackish head and bill

Nest: no nest or minimal nest, on a cliff or in a cave, sometimes in a hollow tree; 1 brood per year

Eggs: 1–3; white with brown markings

Incubation: 38–41 days; female and male incubate

Fledging: 66–88 days; female and male feed the young

Migration: non-migrator to partial migrator, to southern states, Mexico and Central and South America

Food: carrion; parents regurgitate to feed the young

Compare: Black Vulture (p. 41) has shorter wings and tail. Bald Eagle (p. 71) is larger and lacks two-toned wings. Look for the obvious naked red head to identify the Turkey Vulture.

Stan's Notes: The naked head reduces the risk of feather fouling (picking up diseases) from contact with carcasses. It has a strong bill for tearing apart flesh. Unlike hawks and eagles, it has weak feet more suited for walking than grasping. One of the few birds with a developed sense of smell. Mostly mute, making only grunts and groans. Holds its wings in an upright V shape in flight. Teeters from wing tip to wing tip as it soars and hovers. Seen in trees with wings outstretched, sunning itself and drying after a rain.

43

in flight

juvenile

crests

drying

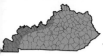

Double-crested Cormorant
Phalacrocorax auritus

Size: 31–35" (79–89 cm); up to 4⅓' wingspan

Male: Large black waterbird with unusual blue eyes and a long, snake-like neck. Large gray bill, with yellow at the base and a hooked tip.

Female: same as male

Juvenile: lighter brown with a grayish chest and neck

Nest: platform; male and female construct; 1 brood per year

Eggs: 3–4; bluish white without markings

Incubation: 25–29 days; female and male incubate

Fledging: 37–42 days; male and female feed the young

Migration: complete, to southern states, Mexico and Central America

Food: small fish, aquatic insects

Compare: The Turkey Vulture (p. 43) is similar in size and also perches on branches with wings open to dry in sun, but it has a naked red head. American Coot (p. 37) lacks the long neck and long pointed bill.

Stan's Notes: Flocks fly in a large V or a line. Usually roosts in large colonies in trees close to water. Swims underwater to catch fish, holding its wings at its sides. This bird's outer feathers soak up water, but its body feathers don't. To dry off, it strikes an upright pose with wings outstretched, facing the sun. Gives grunts, pops and groans. Named "Double-crested" for the crests on its head, which are not often seen. "Cormorant" is a contraction from *corvus marinus*, meaning "crow" or "raven," and "of the sea."

male

female

Black-and-white Warbler
Mniotilta varia

Size: 5" (13 cm)

Male: Small with zebra-like striping and a black-and-white striped crown. Black cheek patch and chin. White belly.

Female: duller than male and lacks a black cheek patch and chin

Juvenile: similar to female

Nest: cup; female builds; 1 brood per year

Eggs: 4–5; white with brown markings

Incubation: 10–11 days; female incubates

Fledging: 9–12 days; female and male feed the young

Migration: complete, to southern states, Mexico and Central and South America

Food: insects

Compare: Climbs down tree trunks headfirst, like the White-breasted Nuthatch (p. 207) and Red-breasted Nuthatch (p. 203). Brown Creeper (p. 95) is brown and white and has a longer, downward-curving bill, and it climbs up trees. Look for a small black-and-white bird climbing down trees to identify the Black-and-white Warbler.

Stan's Notes: The only warbler that moves headfirst down a tree trunk. Look for it searching for insect eggs in the bark of large trees. Its song sounds like a slowly turning, squeaky wheel. The female will perform a distraction dance to draw predators away from the nest. Constructs its nest on the ground, concealed beneath dead leaves or at the base of a tree. A common summer resident, nesting throughout Kentucky.

male

female

Downy Woodpecker
Dryobates pubescens

Size: 6" (15 cm)

Male: Small woodpecker with a white belly and black-and-white spotted wings. Red mark on the back of the head and a white stripe down the back. Short black bill.

Female: same as male but lacks the red mark

Juvenile: same as female, some with a red mark near the forehead

Nest: cavity with a round entrance hole; male and female excavate; 1 brood per year

Eggs: 3–5; white without markings

Incubation: 11–12 days; female incubates during the day, male incubates at night

Fledging: 20–25 days; male and female feed the young

Migration: non-migrator; moves around to find food

Food: insects, seeds; visits seed and suet feeders

Compare: The Hairy Woodpecker (p. 55) is larger. Look for the Downy's shorter, thinner bill.

Stan's Notes: Abundant and widespread where trees are present. This is perhaps the most common woodpecker in the U.S. Stiff tail feathers help to brace it like a tripod as it clings to a tree. Like other woodpeckers, it has a long, barbed tongue to pull insects from tiny places. Mates drum on branches or hollow logs to announce territory, which is rarely larger than 5 acres (2 ha). Repeats a high-pitched "peek-peek" call. Nest cavity is wider at the bottom than at the top and is lined with fallen wood chips. Male performs most of the brooding. During winter, it will roost in a cavity. Undulates in flight.

female
p. 131

male

SUMMER
MIGRATION

Rose-breasted Grosbeak
Pheucticus ludovicianus

Size: 7–8" (18–20 cm)

Male: Plump black-and-white bird with a large triangular, rose-colored patch on the breast. Wing linings are rose-red. Large ivory bill.

Female: heavily streaked with obvious white eyebrows and orange-to-yellow wing linings

Juvenile: similar to female

Nest: cup; female and male construct; 1–2 broods per year

Eggs: 3–5; blue-green with brown markings

Incubation: 13–14 days; female and male incubate

Fledging: 9–12 days; female and male feed the young

Migration: complete, to Mexico, Central America and South America

Food: insects, seeds, fruit; comes to seed feeders

Compare: Male is very distinctive, with no look-alikes. Look for the rose breast patch to identify.

Stan's Notes: Both sexes sing, but the male sings much louder and clearer. Sings a rich, robin-like song with a chip note in the tune. "Grosbeak" refers to the thick, strong bill, which is used to crush seeds. Males arrive at the breeding grounds a few days before females. When females arrive, males become territorial and reduce their feeder visits. After fledging, the young visit feeders with the adults. Late to arrive in spring and early to leave in autumn. Often prefers mature deciduous forest for nesting. The rose breast patch varies in size and shape in each male. Makes short flights from tree to tree with rapid wingbeats.

male

female

Yellow-bellied Sapsucker
Sphyrapicus varius

Size: 8–9" (20–23 cm)

Male: Checkered back with a red forehead, crown and chin. Yellow to tan on the chest and belly. White wing patches are seen flashing in flight.

Female: similar to male but with a white chin

Juvenile: similar to female, dull brown and lacks any red marking

Nest: cavity; female and male excavate, often in a live tree; 1 brood per year

Eggs: 5–6; white without markings

Incubation: 12–13 days; female incubates during the day, male incubates at night

Fledging: 25–29 days; female and male feed the young

Migration: complete, to Kentucky, southern states, Mexico and Central America

Food: insects, tree sap; comes to suet feeders

Compare: The Red-headed Woodpecker (p. 57) has an all-red head. Look for the red chin and crown to identify the male Sapsucker, and the white chin and red crown to identify the female.

Stan's Notes: Found in small woods, forests, and suburban and rural areas. Drills rows of holes in trees to bleed the sap. Oozing sap attracts bugs, which it also eats. Defends its sapping sites from other birds that try to drink from the taps. Does not suck sap; rather, it laps the sticky liquid with its long, bristly tongue. A quiet bird, it makes few vocalizations but will meow like a cat. Drums on hollow branches, but unlike other woodpeckers, its rhythm is irregular. Makes short undulating flights with rapid wingbeats.

male

female

YEAR-ROUND

Hairy Woodpecker
Leuconotopicus villosus

Size: 9" (23 cm)

Male: Black-and-white woodpecker with a white belly. Black wings with rows of white spots. White stripe down the back. Long black bill. Red mark on the back of the head.

Female: same as male but lacks the red mark

Juvenile: grayer version of the female

Nest: cavity with an oval entrance hole; female and male excavate; 1 brood per year

Eggs: 3–6; white without markings

Incubation: 11–15 days; female incubates during the day, male incubates at night

Fledging: 28–30 days; male and female feed the young

Migration: non-migrator

Food: insects, nuts, seeds; comes to seed and suet feeders

Compare: Downy Woodpecker (p. 49) is much smaller and has a much shorter bill. Look for Hairy Woodpecker's long bill, nearly equal to the width of its head.

Stan's Notes: A common bird in wooded backyards. Announces its arrival with a sharp chirp before landing on feeders. Responsible for eating many destructive forest insects. Uses its barbed tongue to extract insects from trees. Tiny, bristle-like feathers at the base of the bill protect the nostrils from wood dust. Drums on hollow logs, branches or stovepipes in spring to announce territory. Often prefers to excavate nest cavities in dead or dying trees. Excavates a larger, more-oval-shaped entrance than the round hole of the Downy Woodpecker. Makes short flights from tree to tree.

juvenile

Red-headed Woodpecker
Melanerpes erythrocephalus

Size: 9" (23 cm)

Male: All-red head with a solid black back. White chest, belly and rump. Black wings with large white wing patches seen flashing in flight. Black tail. Gray legs and bill.

Female: same as male

Juvenile: gray brown with white chest, lacks any red

Nest: cavity; male builds with help from female; 1 brood per year

Eggs: 4–5; white without markings

Incubation: 12–13 days; female and male incubate

Fledging: 27–30 days; female and male feed the young

Migration: partial migrator; will move to areas with an abundant supply of nuts

Food: insects, nuts, fruit; visits suet and seed feeders

Compare: No other woodpecker in Kentucky has an all-red head. The Pileated Woodpecker (p. 67) is the only other woodpecker with a solid black back, but it has a partially red head.

Stan's Notes: One of the few non-dimorphic woodpeckers, with males and females that look alike. Bill is strong enough to excavate a nest cavity only in soft, dead trees. Prefers open woodlands or woodland edges with many dead or rotting branches. Nests later than its close relative, the Red-bellied Woodpecker, and will often take its cavity, if vacant. Unlike other woodpeckers, which use nest cavities just once briefly, it may use the same cavity for several years in a row. Often perches on top of dead snags. Stores acorns and other nuts. Gives a shrill, hoarse "churr" call.

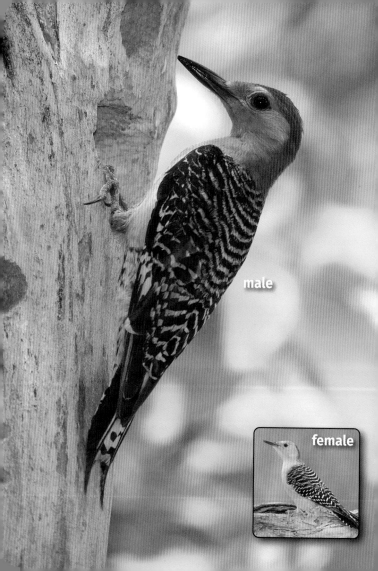

male

female

Red-bellied Woodpecker
Melanerpes carolinus

Size: 9–9½" (23–24 cm)

Male: Black-and-white "zebra-backed" woodpecker with a white rump. Red crown extends down the nape of the neck. Tan chest. Pale-red tinge on the belly, often hard to see.

Female: same as male but with a light-gray crown and a red nape

Juvenile: gray version of adults; lacks a red crown and red nape

Nest: cavity; female and male excavate; 1 brood per year

Eggs: 4–5; white without markings

Incubation: 12–14 days; female incubates during the day, male incubates at night

Fledging: 24–27 days; female and male feed the young

Migration: non-migrator; moves around to find food

Food: insects, nuts, fruit; visits suet and seed feeders

Compare: Similar to the Northern Flicker (p. 157) and Yellow-bellied Sapsucker (p. 53). Look for the zebra-striped back to help identify the Red-bellied Woodpecker.

Stan's Notes: Likes shady woodlands, forest edges and backyards. Digs holes in rotten wood to find spiders, centipedes, beetles and more. Hammers acorns and berries into crevices of trees for winter food. Returns to the same tree to excavate a new nest below that of the previous year. Undulating flight with rapid wingbeats. Gives a loud "querrr" call and a low "chug-chug-chug." Named for the pale red tinge on its belly. Often kicked out of nest hole by European Starlings. Expanding its range all over the country.

female
p. 173

male

Lesser Scaup
Aythya affinis

WINTER

Size: 16–17" (40–43 cm)

Male: Appears mostly black with bold white sides and gray back. Chest and head look nearly black, but head appears purple with green highlights in direct sun. Bright-yellow eyes.

Female: overall brown with dull-white patch at base of light-gray bill, yellow eyes

Juvenile: same as female

Nest: ground; female builds; 1 brood per year

Eggs: 8–14; olive-buff without markings

Incubation: 22–28 days; female incubates

Fledging: 45–50 days; female teaches the young to feed

Migration: complete, to southern states, Mexico and Central America

Food: aquatic plants and insects

Compare: The male Ring-necked Duck (p. 65) has a bold white ring around its bill. The male Blue-winged Teal (p. 167) has a bright-white crescent-shaped mark at the base of bill.

Stan's Notes: A common diving duck in Kentucky. Often in large flocks numbering in the thousands on lakes, ponds and sewage lagoons. Completely submerges itself to feed on the bottom of lakes (unlike dabbling ducks, which only tip forward to reach bottom). Note the bold white stripe under wings when in flight. Quantity of eggs (clutch size) increases with age of female. Interesting baby-sitting arrangement in which groups of young (crèches) are tended by one to three adult females.

female
p. 177

male

Hooded Merganser
Lophodytes cucullatus

YEAR-ROUND

Size: 16–19" (40–48 cm)

Male: Black and white with rust-brown sides. Crest "hood" raises to show a large white patch on each side of the head. Long, thin, black bill.

Female: brown and rust with ragged, rust-red "hair" and a long, thin, brown bill

Juvenile: similar to female

Nest: cavity; female lines an old woodpecker cavity or a nest box near water; 1 brood per year

Eggs: 10–12; white without markings

Incubation: 32–33 days; female incubates

Fledging: 71 days; female feeds the young

Migration: non-migrator in Kentucky

Food: small fish, aquatic insects, crustaceans (especially crayfish)

Compare: Male Wood Duck (p. 255) has a green head. The white patch on the head and rust-brown sides distinguish the male Hoodie.

Stan's Notes: A small diving bird of shallow ponds, sloughs, lakes and rivers, usually in small groups. Quick, low flight across the water, with fast wingbeats. Male has a deep, rolling call. Female gives a hoarse quack. Nests in wooded areas. Female will lay some eggs in the nests of other Hooded Mergansers or Wood Ducks, resulting in 20–25 eggs in some nests. Rarely, she shares a nest, sitting with a Wood Duck.

63

female
p. 175

male

WINTER

Ring-necked Duck
Aythya collaris

Size: 16–19" (41–48 cm)

Male: Striking black duck with light-gray-to-white sides. Blue bill with a bold white ring and a thinner ring at the base. Peaked head with a sloped forehead.

Female: brown with darker-brown back and crown, light-brown sides, gray face, white eye-ring, white ring around the bill, and peaked head

Juvenile: similar to female

Nest: ground; female builds; 1 brood per year

Eggs: 8–10; olive-gray to brown without markings

Incubation: 26–27 days; female incubates

Fledging: 49–56 days; female teaches the young to feed

Migration: complete, to southern states, West Indies, Mexico and Central America

Food: aquatic plants and insects

Compare: Male Lesser Scaup (p. 61) is similar in size, but it has a gray back, unlike the black back of the male Ring-necked Duck. Look for the blue bill with a bold white ring to identify the male Ring-necked Duck.

Stan's Notes: A common winter resident in Kentucky. Often seen in larger lakes, usually in small flocks or just pairs. Watch for this diving duck to dive underwater to forage for food. Springs up off the water to take flight. Flattens its crown when diving. Male gives a quick series of grating barks and grunts. Female gives high-pitched peeps. Named "Ring-necked" for its cinnamon collar, which is nearly impossible to see in the field. Also called Ring-billed Duck due to the white ring on its bill.

male

female

Pileated Woodpecker
Dryocopus pileatus

Size: 19" (48 cm)

Male: Crow-size woodpecker with a black back and bright-red forehead, crest and mustache. Long gray bill. White leading edge of wings flashes brightly during flight.

Female: same as male but with a black forehead; lacks a red mustache

Juvenile: similar to adults but duller and browner

Nest: cavity; male and female excavate; 1 brood per year

Eggs: 3–5; white without markings

Incubation: 15–18 days; female incubates during the day, male incubates at night

Fledging: 26–28 days; female and male feed the young

Migration: non-migrator; moves around to find food

Food: insects; will come to suet and peanut feeders

Compare: The Red-headed Woodpecker (p. 57) is about half the size and has an all-red head. Look for the bright-red crest and exceptionally large size to identify the Pileated Woodpecker.

Stan's Notes: Our largest woodpecker. The common name comes from the Latin *pileatus*, which means "wearing a cap." A relatively shy bird that prefers large tracts of woodland. Drums on hollow branches, chimneys and so forth to announce its territory. Excavates oval holes up to several feet long in tree trunks, looking for insects to eat. Large wood chips lie on the ground by excavated trees. Favorite food is carpenter ants. Feeds regurgitated insects to its young. Young emerge from the nest looking just like the adults.

soaring

SUMMER
MIGRATION

Osprey
Pandion haliaetus

Size: 21–24" (53–61 cm); up to 5½' wingspan

Male: Large eagle-like bird with a white chest, belly and head. Dark eye line. Nearly black back. Black "wrist" marks on the wings. Dark bill.

Female: same as male but slightly larger and with a necklace of brown streaks

Juvenile: similar to adults, with a light-tan breast

Nest: platform on a raised wooden platform, man-made tower or tall dead tree; female and male build; 1 brood per year

Eggs: 2–4; white with brown markings

Incubation: 32–42 days; female and male incubate

Fledging: 48–58 days; male and female feed the young

Migration: complete, to southern states, Mexico and Central and South America

Food: fish

Compare: The juvenile Bald Eagle (p. 71) is brown with white speckles. The adult Bald Eagle has an all-white head and tail. Look for the white belly and dark eye line to identify the Osprey.

Stan's Notes: The only species in its family, and the only raptor that plunges into water feetfirst to catch fish. Always near water. Can hover for a few seconds before diving. Carries fish in a head-first position for better aerodynamics. Wings angle back in flight. Often harassed by Bald Eagles for its catch. Gives a high-pitched, whistle-like call, often calling in flight as a warning. Mates have a long-term pair bond. May not migrate to the same wintering grounds. Was nearly extinct in many areas but is now doing well.

soaring

juvenile

soaring
juvenile

Bald Eagle

Haliaeetus leucocephalus

YEAR-ROUND WINTER

Size: 31–37" (79–94 cm); up to 7½' wingspan

Male: White head and tail contrast sharply with the dark-brown-to-black body and wings. Large, curved yellow bill and yellow feet.

Female: same as male but larger

Juvenile: dark brown with white speckles and spots on the body and wings; gray bill

Nest: massive platform, usually in a tree; female and male build; 1 brood per year

Eggs: 2–3; off-white without markings

Incubation: 34–36 days; female and male incubate

Fledging: 75–90 days; female and male feed the young

Migration: non-migrator to partial migrator, to southeastern states

Food: fish, carrion, birds (mainly ducks)

Compare: Black Vulture (p. 41) is smaller, has a shorter tail and lacks the adult Bald Eagle's white head and tail. Turkey Vulture (p. 43) is smaller and flies with its two-toned wings held in a V shape, unlike the straight-out wing position of the Bald Eagle.

Stan's Notes: Returns to the same nest each year, adding more sticks and enlarging it to huge proportions, at times up to 1,000 pounds (450 kg). In their midair mating ritual, one eagle flips upside down and locks talons with another. Both tumble, then break apart to continue flight. Not uncommon for juveniles to perform this mating ritual even though they are not yet breeding age. Long-term pair bond but will switch mates when not successful at reproducing. Juveniles attain the white head and tail at 4–5 years of age.

Blue-gray Gnatcatcher
Polioptila caerulea

Size: 4" (10 cm)

Male: A light-blue-to-gray head, back, breast and wings, with a white belly. Black forehead and eyebrows. Prominent white eye-ring. Long black tail with a white undertail, often held cocked above the rest of body.

Female: same as male but grayer and lacking black on the head

Juvenile: similar to female

Nest: cup; female and male construct; 1 brood per year

Eggs: 4–5; pale blue with dark markings

Incubation: 10–13 days; female and male incubate

Fledging: 10–12 days; female and male feed the young

Migration: complete, to southern states, the Bahamas, Mexico and Central America

Food: insects

Compare: The only small blue bird with a black tail. Very active near the nest. Look for it flitting around upper branches in search of insects.

Stan's Notes: Found in a wide variety of forest types throughout Kentucky. Listen for its wheezy call notes to help locate it. A fun and easy bird to watch. Flicks its tail up and down and from side to side while calling. In many years, it nests so early that by mid-June it is no longer defending territory, with most leaving by the end of August. Like many open-woodland nesters, it is a common cowbird host. Although the population is abundant and widespread, it has been decreasing in the recent past.

female
p. 107

male

SUMMER

Indigo Bunting
Passerina cyanea

Size: 5½" (14 cm)

Male: Vibrant-blue finch-like bird. Dark markings scattered on wings and tail.

Female: light-brown with faint markings

Juvenile: similar to female

Nest: cup; female builds; 2 broods per year

Eggs: 3–4; pale blue without markings

Incubation: 12–13 days; female incubates

Fledging: 10–11 days; female feeds the young

Migration: complete, to southern states, Mexico and Central America

Food: insects, seeds, fruit; will visit seed feeders

Compare: The male Eastern Bluebird (p. 83) is larger and has a rust-red chest. Male Blue Grosbeak (p. 81) is larger, has chestnut-colored wing bars and a large bill.

Stan's Notes: Seen along woodland edges and in parks and yards, feeding on insects. Comes to seed feeders early in spring, before insects are plentiful. Usually only the males are noticed. The male often sings from treetops to attract a mate. The female is quiet. Actually a gray bird, without blue pigment in its feathers: like Blue Jays and other blue birds, sunlight is refracted within the structure of the feathers, making them appear blue. Plumage is iridescent in direct sun, duller in shade. Molts in spring to acquire body feathers with gray tips, which quickly wear off, revealing the bright-blue plumage. Molts in fall and appears like the female during winter. Migrates at night in flocks of 5–10 birds. Males return before the females and juveniles, often to the nest site of the preceding year. Juveniles move to within a mile of their birth site.

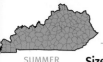

SUMMER

Tree Swallow
Tachycineta bicolor

Size: 5–6" (13–15 cm)

Male: Blue-green in spring, greener in fall. Changes color in direct sunlight. White from chin to belly. Long, pointed wing tips. Notched tail.

Female: similar to male but duller

Juvenile: gray brown with a white belly and a grayish breast band

Nest: cavity; female and male line a vacant wood-pecker cavity or nest box; 2 broods per year

Eggs: 4–6; white without markings

Incubation: 13–16 days; female incubates

Fledging: 20–24 days; female and male feed the young

Migration: complete, to Mexico and Central America

Food: insects

Compare: The Purple Martin (p. 85) is much larger and darker. The Barn Swallow (p. 79) has a rusty belly and a long, deeply forked tail. Look for the white chin, chest and belly and the notched tail to help identify the Tree Swallow.

Stan's Notes: The first swallow to return each spring. Most common along ponds, lakes and agricultural fields. Can be attracted to your yard with a nest box. Competes with Eastern Bluebirds for cavities and nest boxes. Builds a grass nest within and will travel long distances, looking for dropped feathers for the lining. Flies with rapid wingbeats, then glides. Watch for it playing and chasing after feathers. Gives a series of gurgles and chirps. Chatters when upset or threatened. Eats many nuisance bugs. Families gather in large flocks for migration.

Barn Swallow
Hirundo rustica

SUMMER

Size: 7" (18 cm)

Male: Sleek swallow. Blue-black back, cinnamon belly and reddish-brown chin. White spots on a long, deeply forked tail.

Female: same as male but with a whitish belly

Juvenile: similar to adults, with a tan belly and chin, and shorter tail

Nest: cup; female and male build; 2 broods per year

Eggs: 4–5; white with brown markings

Incubation: 13–17 days; female incubates

Fledging: 18–23 days; female and male feed the young

Migration: complete, to South America

Food: insects (prefers beetles, wasps, flies)

Compare: The Tree Swallow (p. 77) is white from chin to belly. The Purple Martin (p. 85) is larger and has a dark-purple belly. The Chimney Swift (p. 91) has a narrow, pointed tail. Look for the deeply forked tail to identify the Barn Swallow.

Stan's Notes: Seen in wetlands, farms, suburban yards and parks. Of the six swallow species in Kentucky, this is the only one with a deeply forked tail. Unlike other swallows, it rarely glides in flight. Usually flies low over land or water. Drinks as it flies, skimming water, or will sip water droplets on wet leaves. Bathes while flying through rain or sprinklers. Gives a twittering warble, followed by a mechanical sound. Builds a mud nest with up to 1,000 beak-loads of mud. Nests on barns and houses, under bridges and in other sheltered places. Often nests in colonies of 4–6 birds; sometimes nests alone.

female
p. 123

male

Blue Grosbeak
Passerina caerulea

SUMMER MIGRATION

Size: 7" (18 cm)

Male: Overall blue bird with 2 chestnut wing bars. Large gray-to-silver bill. Black around base of bill.

Female: overall brown with darker wings and tail, 2 tan wing bars, large gray-to-silver bill

Juvenile: similar to female

Nest: cup; female builds; 1–2 broods per year

Eggs: 3–6; pale blue without markings

Incubation: 11–12 days; female incubates

Fledging: 9–10 days; female and male feed the young

Migration: complete, to Florida, Mexico and Central America

Food: insects, seeds; will come to seed feeders

Compare: The more common male Indigo Bunting (p. 75) is very similar, but it is smaller and lacks wing bars. The male Eastern Bluebird (p. 83) is the same size, but it lacks the wing bars and oversized bill.

Stan's Notes: Increasing population in Kentucky over the past 30 to 40 years, Blue Grosbeaks return to the state by mid-April. A bird of semi-open habitats such as overgrown fields, riversides, woodland edges and fencerows. Visits seed feeders, where it is often confused with male Indigo Buntings. Frequently seen twitching and spreading its tail. The first-year males show only some blue, obtaining the full complement of blue feathers in the second winter.

male

female

Eastern Bluebird
Sialia sialis

YEAR-ROUND

Size: 7" (18 cm)

Male: Sky-blue head, back and tail. Rust-red breast and white belly.

Female: grayer than male, with a faint rusty breast and faint blue wings and tail

Juvenile: similar to female but with spots on the breast and blue wing markings

Nest: cavity, vacant woodpecker cavity or nest box; female adds a soft lining; 2 broods per year

Eggs: 4–5; pale blue without markings

Incubation: 12–14 days; female incubates

Fledging: 15–18 days; male and female feed the young

Migration: non-migrator to partial; moves around to find food

Food: insects, fruit; comes to shallow dishes with live or dead mealworms, and to suet feeders

Compare: The male Indigo Bunting (p. 75) is nearly all blue. The Blue Jay (p. 87) is much larger and has a crest. Look for the rusty breast to help identify the Eastern Bluebird.

Stan's Notes: A year-round resident in Kentucky, Eastern Bluebird populations sometimes drop due to unseasonably cold winters or cold, wet weather in spring. Although a permanent resident, some Eastern Bluebirds migrate each spring and autumn. Prefers open habitats, such as farm fields, pastures and roadsides, but also likes forest edges, parks and yards. Often perches on trees or fence posts and drops to the ground to grab bugs. Song is a distinctive "churlee chur chur-lee." The rust-red breast is like that of the American Robin, its cousin.

83

male

female

Purple Martin
Progne subis

SUMMER

Size: 8½" (22 cm)

Male: Iridescent with a purple-to-black head, back and belly. Black wings and a notched black tail.

Female: grayish-purple head and back, darker wings and tail, whitish belly

Juvenile: same as female

Nest: cavity; female and male line the cavity of the house; 1 brood per year

Eggs: 4–5; white without markings

Incubation: 15–18 days; female incubates

Fledging: 26–30 days; male and female feed the young

Migration: complete, to South America

Food: insects

Compare: Usually seen only in groups. The male Purple Martin is the only swallow with a very dark-purplish belly.

Stan's Notes: The largest swallow species in North America. Once nested in tree cavities; now nests almost exclusively in man-made, apartment-style houses. The most successful colonies often nest in multiunit nest boxes within 100 feet (30 m) of a human dwelling near a lake. Main diet consists of dragonflies, not mosquitoes, as once thought. Gives a continuous stream of chirps, creaks and rattles, along with a shout-like "churrr" and chortle. Often drinks in flight, skimming water, and bathes in flight, flying through rain. Returns to the same nest site each year; the males arrive before the females and yearlings. The young leave to form new colonies. Large colonies gather in fall before migrating to South America.

Blue Jay
Cyanocitta cristata

YEAR-ROUND

Size: 12" (30 cm)

Male: Bright light-blue-and-white bird with a black necklace and gray belly. Large crest moves up and down at will. White face, wing bars and tip of tail. Black tail bands.

Female: same as male

Juvenile: same as adult but duller

Nest: cup; female and male construct; 1–2 broods per year

Eggs: 4–5; green to blue with brown markings

Incubation: 16–18 days; female incubates

Fledging: 17–21 days; female and male feed the young

Migration: non-migrator to partial migrator; will move around to find an abundant food source

Food: insects, fruit, carrion, seeds, nuts; visits seed feeders, ground feeders with corn or peanuts

Compare: The Belted Kingfisher (p. 89) has a larger, more ragged crest. The Eastern Bluebird (p. 83) is much smaller and has a rust-red breast. Look for the large crest to help identify the Blue Jay.

Stan's Notes: Highly intelligent, solving problems, gathering food and communicating more than other birds. Loud and noisy; mimics other birds. Known as the alarm of the forest, screaming at intruders. Imitates hawk calls around feeders to scare off other birds. One of the few birds to cache food; can remember where it hid thousands of nuts. Carries food in a pouch under its tongue (sublingually). Eats eggs and young birds from other nests. Feathers lack blue pigment; refracted sunlight causes the blue appearance.

YEAR-ROUND

Belted Kingfisher
Megaceryle alcyon

Size: 12–14" (30–36 cm)

Male: Blue with white belly, blue-gray chest band, and black wing tips. Ragged crest moves up and down at will. Large head. Long, thick, black bill. White spot by eyes. Red-brown eyes.

Female: same as male but with rusty flanks and a rusty chest band below the blue-gray band

Juvenile: similar to female

Nest: cavity; female and male excavate in a bank of a river, lake or cliff; 1 brood per year

Eggs: 6–7; white without markings

Incubation: 23–24 days; female and male incubate

Fledging: 23–24 days; female and male feed the young

Migration: non-migrator to partial; moves around to find food

Food: small fish

Compare: The Blue Jay (p. 87) is lighter blue and has a plain gray chest and belly. The Belted Kingfisher is rarely found away from water.

Stan's Notes: Usually found at the bank of a river, lake or large stream. Perches on a branch near water, dives in headfirst to catch a small fish, then returns to the branch to feed. Parents drop dead fish into the water to teach their young to dive. Can't pass bones through its digestive tract; regurgitates bone pellets after meals. Gives a loud call that sounds like a machine gun. Mates know each other by their calls. Digs a tunnel up to 4 feet (1 m) long to a nest chamber. Small white patches on dark wing tips flash during flight.

Chimney Swift
Chaetura pelagica

Size: 5" (13 cm)

Male: Nondescript, cigar-shaped bird, usually seen in flight. Long, thin, brown body. Pointed tail and head. Long, backswept wings, longer than the body.

Female: same as male

Juvenile: same as adult

Nest: half cup; female and male build; 1 brood per year

Eggs: 4–5; white without markings

Incubation: 19–21 days; female and male incubate

Fledging: 28–30 days; female and male feed the young

Migration: complete, to South America

Food: insects caught in midair

Compare: The Purple Martin (p. 85) is much larger and darker. The Barn Swallow (p. 79) has a deeply forked tail. Tree Swallows (p. 77) have a white belly and blue-green back. Look for the cigar shape to identify the Chimney Swift in flight.

Stan's Notes: One of the fastest fliers in the bird world. Spends all day flying, rarely perching. Flies in groups, feeding on insects flying 100 feet (30 m) or higher up in the air. Often called a Flying Cigar due to its body shape, which is pointed at both ends. Drinks and bathes during flight, skimming water. Gives a unique in-flight twittering call, often heard before the bird is seen. Hundreds roost in large chimneys, giving it the common name. Builds its nest with tiny twigs, cementing it with saliva and attaching it to the inside of a chimney or a hollow tree. Usually only one nest per chimney.

SUMMER
MIGRATION

Chestnut-sided Warbler
Setophaga pensylvanica

Size: 5" (13 cm)

Male: Colorful combination of a bright-yellow cap, black mask and white face, with a white chest and belly. Chestnut flanks. Gray wings with 2 yellow wing bars. White undertail.

Female: similar to male, with duller brown flanks

Juvenile: similar to female, with a lime-green head and back, white eye-ring and bright-yellow wing bars; lacks chestnut sides

Nest: cup; female builds; 1 brood per year

Eggs: 3–5; white with brown markings

Incubation: 12–13 days; female incubates

Fledging: 10–12 days; female and male feed the young

Migration: complete, to Mexico and Central America

Food: insects, berries

Compare: The Yellow-rumped Warbler (p. 209) has yellow patches on its sides and rump. Yellow Warbler (p. 283) is nearly all yellow. Look for the yellow cap and chestnut flanks to help identify the Chestnut-sided Warbler.

Stan's Notes: A very attractive warbler, named for the chestnut patches on its sides. During migration, it is often attracted to backyard water gardens that have a small trickling stream. Look for it in spring, hopping high up in trees while it hunts for insects. Usually you will get only a glimpse of this fast-moving bird. Holds tail in an uplifted position, showing the white undertail. It is not uncommon for it to approach people near its nest in defense of the site.

Brown Creeper
Certhia americana

YEAR-ROUND
WINTER

Size: 5" (13 cm)

Male: Small, thin, nearly camouflaged brown bird. White from chin to belly. White eyebrows. Dark eyes and a thin, curved bill. Tail is long and stiff.

Female: same as male

Juvenile: same as adults

Nest: cup; female constructs; 1 brood per year

Eggs: 5–6; white with tiny brown markings

Incubation: 14–17 days; female incubates; male feeds the female during incubation

Fledging: 13–16 days; female and male feed the young

Migration: complete to partial migrator

Food: insects, nuts, seeds

Compare: The Red-breasted Nuthatch (p. 203) and White-breasted Nuthatch (p. 207) climb down tree trunks, not up. To spot a Brown Creeper, look for a small brown bird with a white belly creeping up trees.

Stan's Notes: A forest bird, commonly found in wooded habitats. Will fly from the top of one tree trunk to the bottom of another, then work its way to the top, looking for caterpillars, spider eggs and more. Its long tail has tiny spines underneath, which help it cling to trees. Uses its camouflage coloring to hide in plain sight: it spreads out flat on a branch or trunk and won't move. Often builds its nest behind the loose bark of a dead or dying tree. Young follow their parents around, creeping up trees soon after fledging.

Chipping Sparrow
Spizella passerina

Size: 5" (13 cm)

Male: Small gray-brown sparrow with clear-gray chest. Rusty crown. White eyebrows and thin black eye line. Thin gray-black bill. Two faint wing bars.

Female: same as male

Juvenile: similar to adults, with streaking on the chest; lacks a rusty crown

Nest: cup; female builds; 2 broods per year

Eggs: 3–5; blue-green with brown markings

Incubation: 11–14 days; female incubates

Fledging: 10–12 days; female and male feed the young

Migration: complete, to southern coastal states, Mexico and Central America

Food: insects, seeds; will come to ground feeders

Compare: The rusty Fox Sparrow (p. 121) is larger and lacks a clear breast. The Song Sparrow (p. 111) and female House Finch (p. 101) have heavily streaked chests. Look for the rusty crown and black eye line to help identify the Chipping Sparrow.

Stan's Notes: A common garden or yard bird, often seen feeding on dropped seeds beneath feeders. Gathers in large family groups to feed in preparation for migration. Migrates at night in flocks of 20–30 birds. The common name comes from the male's fast "chip" call. Often is just called Chippy. Builds nest low in dense shrubs and almost always lines it with animal hair. Comfortable with people, allowing you to approach closely before it flies away.

Pine Siskin
Spinus pinus

WINTER

Size: 5" (13 cm)

Male: Small brown finch with heavy streaking on the back, breast and belly. Yellow wing bars. Yellow at the base of the tail. Thin bill.

Female: similar to male, with less yellow

Juvenile: similar to adult, with a light-yellow tinge over the breast and chin

Nest: cup; female builds; 2 broods per year

Eggs: 3–4; greenish blue with brown markings

Incubation: 12–13 days; female incubates

Fledging: 14–15 days; female and male feed the young

Migration: irruptive; moves around the United States in search of food during winter

Food: seeds, insects; will come to seed feeders

Compare: The female Purple Finch (p. 115) has white eyebrows. The female House Finch (p. 101) lacks yellow. The female American Goldfinch (p. 295) has white wing bars. Look for the yellow wing bars to identify the Pine Siskin.

Stan's Notes: Usually considered a winter finch, this bird can be seen in flocks of up to 20 birds, often with other finch species. Found throughout the state in heavy invasion years; it is absent from the state in some winters. Gathers in flocks and moves around, visiting feeders. Will come to thistle feeders. Gives a series of high-pitched, wheezy calls. Also gives a wheezing twitter. Breeds in small groups. Builds nest toward the end of coniferous branches, where needles are dense, helping to conceal. Nests are often only a few feet apart. Male feeds the female during incubation. Juveniles lose the yellow tint by late summer of their first year.

male
p. 267

female

House Finch
Haemorhous mexicanus

YEAR-ROUND

Size: 5" (13 cm)

Female: Plain brown with heavy streaking on a white chest.

Male: red-to-orange face, throat, chest and rump; streaked belly and wings; brown cap; brown marking behind the eyes

Juvenile: similar to female

Nest: cup, occasionally in a cavity; female builds; 2 broods per year

Eggs: 4–5; pale blue, lightly marked

Incubation: 12–14 days; female incubates

Fledging: 15–19 days; female and male feed the young

Migration: non-migrator to partial migrator; will move around to find food

Food: seeds, fruit, leaf buds; visits seed feeders and feeders that offer grape jelly

Compare: The female Purple Finch (p. 115) has bold white eyebrows. The Pine Siskin (p. 99) has yellow wing bars and a smaller bill. The female American Goldfinch (p. 295) has a clear chest. Look for the heavily streaked chest to help identify the female House Finch.

Stan's Notes: A relatively new bird to Kentucky (first reported in the early 1970s and first nests reported by 1981). A very social bird, visiting feeders in small flocks. Likes to nest in hanging flower baskets. Male sings a loud, cheerful warbling song. It was originally introduced to Long Island, New York, from the western U.S. in the 1940s. Now found throughout the country. Suffers from a disease that causes the eyes to crust, resulting in blindness and death.

SUMMER

House Wren
Troglodytes aedon

Size: 5" (13 cm)

Male: All-brown bird with lighter-brown markings on the wings and tail. Slightly curved brown bill. Often holds tail upward.

Female: same as male

Juvenile: same as adult

Nest: cavity; female and male line just about any nest cavity; 2 broods per year

Eggs: 4–6; tan with brown markings

Incubation: 10–13 days; female and male incubate

Fledging: 12–15 days; female and male feed the young

Migration: complete, to southern states and Mexico

Food: insects, spiders, snails

Compare: Carolina Wren (p. 105) has prominent eyebrows. Look for House Wren's long curved bill and long upturned tail to differentiate it from sparrows.

Stan's Notes: A prolific songster. During the mating season, sings from dawn to dusk. Seen in brushy yards, parks and woodlands and along forest edges. Easily attracted to a nest box. In spring, the male chooses several prospective nesting cavities and places a few small twigs in each. The female inspects all of them and finishes constructing the nest in the cavity of her choice. She fills the cavity with short twigs and then lines a small depression at the back with pine needles and grass. She often has trouble fitting longer twigs through the entrance hole and tries many different directions and approaches until she is successful.

Carolina Wren
Thryothorus ludovicianus

Size: 5½" (14 cm)

Male: Rusty-brown head and back with an orange-yellow chest and belly. White throat and a prominent white eye stripe. Short, stubby tail, often cocked up.

Female: same as male

Juvenile: same as adults

Nest: cavity; female and male build; 2 broods per year, sometimes 3

Eggs: 4–6; white, sometimes pink or creamy, with brown markings

Incubation: 12–14 days; female incubates

Fledging: 12–14 days; female and male feed the young

Migration: non-migrator; moves around in winter

Food: insects, fruit, few seeds; visits suet feeders

Compare: House Wren (p. 103) is darker brown and lacks a white eye stripe.

Stan's Notes: Mates are long-term, staying together throughout the year in permanent territories. Sings year-round. The male is known to sing up to 40 song types, singing one song repeatedly before switching to another. The female also sings, resulting in duets. The male often takes over feeding the first brood while the female renests. Nests in birdhouses and in unusual places like mailboxes, bumpers or broken taillights of vehicles, or nearly any other cavity. Found in brushy yards or woodlands. Can be attracted to feeders with mealworms.

female

male
p. 75

SUMMER

Indigo Bunting
Passerina cyanea

Size: 5½" (14 cm)

Female: Light-brown, finch-like bird. Faint streaking on a light-tan chest. Wings have a very faint blue cast and indistinct wing bars.

Male: vibrant blue with scattered dark markings on wings and tail

Juvenile: similar to female

Nest: cup; female builds; 2 broods per year

Eggs: 3–4; pale blue without markings

Incubation: 12–13 days; female incubates

Fledging: 10–11 days; female feeds the young

Migration: complete, to southern states, Mexico and Central America

Food: insects, seeds, fruit; will visit seed feeders

Compare: Female Blue Grosbeak (p. 123) is larger and has 2 tan wing bars. The female Purple Finch (p. 115) has white eyebrows and heavy streaking on the chest. The female House Finch (p. 101) has a heavily streaked chest. The female American Goldfinch (p. 295) has white wing bars. Look for the faint blue cast on the wings to help identify the female Indigo Bunting.

Stan's Notes: Seen along woodland edges and in parks and yards, feeding on insects. Comes to seed feeders early in spring, before insects are plentiful. Secretive, plain and quiet; usually only the males are noticed. The male often sings from treetops to attract a mate. Migrates at night in flocks of 5–10 birds. Males return before the females and juveniles, often to the nest site of the preceding year. Juveniles move to within a mile of their birth site.

male
p. 213

female

Dark-eyed Junco
Junco hyemalis

YEAR-ROUND
WINTER

Size: 5½" (14 cm)

Female: Plump, dark-eyed bird with a tan-to-brown chest, head and back. White belly. Ivory-to-pink bill. White outer tail feathers appear like a white V in flight.

Male: round with gray plumage

Juvenile: similar to female, with streaking on the breast and head

Nest: cup; female and male build; 2 broods per year

Eggs: 3–5; white with reddish-brown markings

Incubation: 12–13 days; female incubates

Fledging: 10–13 days; male and female feed the young

Migration: complete, throughout the U.S.; winters in Kentucky; non-migrator in parts of eastern Kentucky

Food: seeds, insects; visits ground and seed feeders

Compare: Rarely confused with any other bird. Look for the ivory-to-pink bill and small flocks feeding under feeders to help identify the female Dark-eyed Junco.

Stan's Notes: A common winter bird of Kentucky. Migrates from Canada to Kentucky and beyond. Females tend to migrate farther south than males. Adheres to a rigid social hierarchy, with dominant birds chasing the less dominant ones. Look for the white outer tail feathers flashing in flight. Often seen in small flocks on the ground, where it uses its feet to simultaneously "double-scratch" to expose seeds and insects. Several subspecies of Dark-eyed Junco were previously considered to be separate species.

Song Sparrow
Melospiza melodia

Size: 5–6" (13–15 cm)

Male: Common brown sparrow with heavy dark streaks on the chest coalescing into a central dark spot.

Female: same as male

Juvenile: similar to adults, with a finely streaked chest; lacks a central dark spot

Nest: cup; female builds; 2 broods per year

Eggs: 3–4; blue to green, with red-brown markings

Incubation: 12–14 days; female incubates

Fledging: 9–12 days; female and male feed the young

Migration: non-migrator in Kentucky; moves around to find food

Food: insects, seeds; only rarely comes to ground feeders with seeds

Compare: Similar to other brown sparrows. Look for the heavily streaked chest with a central dark spot to help identify the Song Sparrow.

Stan's Notes: There are many subspecies of this bird, but the dark spot in the center of the chest appears in every variety. A constant songster, repeating its loud, clear song every few minutes. The song varies from region to region but has the same basic structure. Sings from thick shrubs to defend a small territory, beginning with three notes and finishing up with a trill. A ground feeder, it will "double-scratch" with both feet at the same time to expose seeds. When the female builds a new nest for a second brood, the male often takes over feeding the first brood. Unlike many other sparrow species, Song Sparrows rarely flock together. A common host of the Brown-headed Cowbird.

male

female

House Sparrow
Passer domesticus

YEAR-ROUND

Size: 6" (15 cm)

Male: Brown back with a gray belly and cap. Large black patch extending from the throat to the chest (bib). One white wing bar.

Female: slightly smaller than the male; light brown with light eyebrows; lacks a bib and white wing bar

Juvenile: similar to female

Nest: cavity; female and male build a domed cup nest within; 2–3 broods per year

Eggs: 4–6; white with brown markings

Incubation: 10–12 days; female incubates

Fledging: 14–17 days; female and male feed the young

Migration: non-migrator; moves around to find food

Food: seeds, insects, fruit; comes to seed feeders

Compare: Chipping Sparrow (p. 97) has a rusty crown. Look for the black bib to identify the male House Sparrow and the clear breast to help identify the female.

Stan's Notes: One of the first birdsongs heard in cities in spring. A familiar city bird, nearly always in small flocks. Also found on farms. Introduced in 1850 from Europe to Central Park in New York. Now seen throughout North America. Related to old-world sparrows; not a relative of any sparrows in the U.S. An aggressive bird that will kill young birds in order to take over the nest cavity. Uses dried grass and small scraps of plastic, paper and other materials to build an oversize, domed nest in the cavity.

male
p. 269

female

Purple Finch
Haemorhous purpureus

Size: 6" (15 cm)

Female: Plain brown with heavy streaking on the chest. Bold white eyebrows and a large bill.

Male: raspberry-red head, cap, breast, back and rump; brownish wings and tail

Juvenile: same as female

Nest: cup; female and male build; 1 brood per year

Eggs: 4–5; greenish blue with brown markings

Incubation: 12–13 days; female incubates

Fledging: 13–14 days; female and male feed the young

Migration: irruptive; moves around in search of food

Food: seeds, insects, fruit; comes to seed feeders

Compare: The female House Finch (p. 101) lacks eyebrows. Pine Siskin (p. 99) has yellow wing bars. The female American Goldfinch (p. 295) has a clear chest. Look for the bold white eyebrows to identify the female Purple Finch.

Stan's Notes: Usually seen only during the winter, when flocks of Purple Finches leave their homes farther north and move around searching for food. Travels in flocks of up to 50 birds. Visits seed feeders along with House Finches, which makes it hard to tell them apart. Feeds mainly on seeds; ash tree seeds are an important source of food. Found in coniferous forests, mixed woods, woodland edges and suburban backyards. Flies in the typical undulating, up-and-down pattern of finches. Sings a rich, loud song. Gives a distinctive "tic" note only in flight. The male is not purple. The Latin species name *purpureus* means "purple" (and other reddish colors).

white-striped

tan-striped

White-throated Sparrow
Zonotrichia albicollis

Size: 6–7" (15–18 cm)

Male: Brown with a gray or tan chest and belly. Bold striping on the head. White or tan throat patch and eyebrows. Small yellow spot in the space between the eye and bill, called the lore.

Female: same as male

Juvenile: similar to adults, with a heavily streaked chest and a gray throat and eyebrows

Nest: cup; female builds; 1 brood per year

Eggs: 4–6; green to blue, or creamy white with red-brown markings

Incubation: 11–14 days; female incubates

Fledging: 10–12 days; female and male feed the young

Migration: complete, to Kentucky, southern states and Mexico

Food: insects, seeds, fruit; visits ground feeders

Compare: The White-crowned Sparrow (p. 119) lacks the throat patch and yellow lores of the White-throated Sparrow. Song Sparrow (p. 111) has a central spot on chest and lacks a striped pattern on head.

Stan's Notes: Two color variations (polymorphic): white-striped and tan-striped. Studies indicate that the white-striped adults tend to mate with the tan-striped birds; it's not clear why. Known for its wonderful song; it sings all year and can even be heard at night. White- and tan-striped males and white-striped females sing, but tan-striped females do not. A winter resident that can be seen at ground feeders.

juvenile

WINTER

White-crowned Sparrow
Zonotrichia leucophrys

Size: 6½–7½" (16.5–19 cm)

Male: Brown with a gray chest and black-and-white striped crown. Small, thin pink bill.

Female: same as male

Juvenile: similar to adults, with black and brown stripes on the head

Nest: cup; female builds; 2 broods per year

Eggs: 3–5; greenish to bluish to whitish with red-brown markings

Incubation: 11–14 days; female incubates

Fledging: 8–12 days; male and female feed the young

Migration: complete, to Kentucky, southern states and Mexico

Food: insects, seeds, berries; visits ground feeders

Compare: The White-throated Sparrow (p. 117) has a throat patch and a small yellow spot by each eye. The Song Sparrow (p. 111) has a streaked chest. Look for the striped crown to help identify the White-crowned Sparrow.

Stan's Notes: Winter visitor throughout Kentucky. Often seen in groups of up to 20 birds during migration, when it can be seen visiting ground feeders and feeding beneath seed feeders. This ground feeder will "double-scratch" backward with both feet simultaneously to find seeds. Prefers scrubby areas, woodland edges, and open or grassy habitats. Males arrive at the breeding grounds before the females and sing from perches to establish territory. Males take most of the responsibility for raising the young while females start their second broods. Only 9–12 days separate the broods. Nests in Canada and Alaska.

Fox Sparrow
Passerella iliaca

WINTER

Size: 7" (18 cm)

Male: Plump, rust-red sparrow. Heavily streaked rusty breast and solid-rust tail. Head and back are mottled with gray.

Female: same as male

Juvenile: same as adults

Nest: cup; female builds; 2 broods per year

Eggs: 2–4; pale green with reddish markings

Incubation: 12–14 days; female incubates

Fledging: 10–11 days; female and male feed the young

Migration: complete, to Kentucky and southern states

Food: seeds, insects; comes to ground feeders

Compare: The Brown Thrasher (p. 155) is much larger, is slimmer and has a long, curved bill. The rust-red plumage of the Fox Sparrow differentiates it from all other sparrows.

Stan's Notes: One of the largest sparrows. Often seen alone or in small groups. Found in shrubby areas, open fields and backyards. Comes to ground feeders; searches for seeds and insects underneath seed feeders during migration. Like a chicken, it will "double-scratch" with both feet at the same time to look for food. Gives a series of rich notes lasting 2–3 seconds, usually singing from a perch hidden in a shrub. The common name "Sparrow" comes from the Anglo-Saxon word *spearwa*, meaning "flutterer," as applies to any small bird. "Fox" refers to its rusty color. Appears in several color variations, depending on location. Nests on the ground in brush and along forest edges in Canada and Alaska.

male
p. 81

female

Blue Grosbeak
Passerina caerulea

Size: 7" (18 cm)

Female: Overall brown with darker wings and tail. Two tan wing bars. Large gray-to-silver bill.

Male: blue bird with 2 chestnut wing bars; large gray-to-silver bill; black around base of bill

Juvenile: similar to female

Nest: cup; female builds; 1–2 broods per year

Eggs: 3–6; pale blue without markings

Incubation: 11–12 days; female incubates

Fledging: 9–10 days; female and male feed the young

Migration: complete, to Florida, Mexico and Central America

Food: insects, seeds; will come to seed feeders

Compare: The female Indigo Bunting (p. 107) is very similar, but it is more common, lacks the tan wing bars and is lighter in color overall.

Stan's Notes: Increasing population in Kentucky over the past 30 to 40 years, Blue Grosbeaks return to the state by mid-April. A bird of semi-open habitats such as overgrown fields, riversides, woodland edges and fencerows. Visits seed feeders, where it can be confused with female Indigo Buntings. Often seen twitching and spreading its tail. The first-year males show only some blue, obtaining the full complement of blue feathers in the second winter.

male
p. 27

female

Brown-headed Cowbird
Molothrus ater

YEAR-ROUND

Size: 7½" (19 cm)

Female: Dull brown with no obvious markings. Pointed, sharp, gray bill. Dark eyes.

Male: glossy black with a chocolate-brown head

Juvenile: similar to female but with dull-gray plumage and a streaked chest

Nest: no nest; lays eggs in the nests of other birds

Eggs: 5–7; white with brown markings

Incubation: 10–13 days; host bird incubates the eggs

Fledging: 10–11 days; host birds feed the young

Migration: non-migrator; moves around to find food

Food: insects, seeds; will come to seed feeders

Compare: The female Red-winged Blackbird (p. 141) has white eyebrows and heavy streaking. The female Indigo Bunting (p. 107) has faint blue on its wings. The pointed gray bill helps to identify the female Brown-headed Cowbird.

Stan's Notes: Cowbirds are members of the blackbird family. Known as brood parasites, Brown-headed Cowbirds are the only parasitic birds in Kentucky. Brood parasites lay their eggs in the nests of other birds, leaving the host birds to raise their young. Cowbirds are known to have laid their eggs in the nests of over 200 species of birds. While some birds reject cowbird eggs, most incubate them and raise the young, even to the exclusion of their own. Look for warblers and other birds feeding young birds twice their own size. Named "Cowbird" for its habit of following bison and cattle herds to feed on insects flushed up by the animals.

1 year
old

Cedar Waxwing
Bombycilla cedrorum

Size: 7½" (19 cm)

Male: Sleek-looking, gray-to-brown bird. Pointed crest, bandit-like mask and light-yellow belly. Bold-yellow tip of tail. Red wing tips look like they were dipped in red wax.

Female: same as male

Juvenile: grayish with a heavily streaked breast; lacks the sleek look, black mask and red wing tips

Nest: cup; female and male construct; 1 brood per year, occasionally 2

Eggs: 4–6; pale blue with brown markings

Incubation: 10–12 days; female incubates

Fledging: 14–18 days; female and male feed the young

Migration: non-migrator to partial migrator; moves around to find food

Food: cedar cones, fruit, insects

Compare: The female Northern Cardinal (p. 139) has a large red bill. Look for the red wing tips to identify the Cedar Waxwing.

Stan's Notes: The name is derived from its red, wax-like wing tips and preference for the small, berry-like cones of the cedar. Seen in flocks, moving around from area to area looking for berries. Feeds on insects during summer, before berries are abundant. Wanders during winter, searching for food supplies. Spends most of its time at the top of tall trees. Listen for the high-pitched "sreee" whistling sound it constantly makes while perched or in flight. Obtains the mask after the first year and red wing tips after the second year.

female

male
p. 25

Eastern Towhee
Pipilo erythrophthalmus

YEAR-ROUND

Size: 7–8" (18–20 cm)

Female: Mostly light-brown bird. Rusty red-brown sides and a white belly. Long brown tail with a white tip. Short, stout, pointed bill and rich, red eyes. White wing patches flash in flight.

Male: similar to female but black instead of brown

Juvenile: light brown with a heavily streaked head, chest and belly; long dark tail with a white tip

Nest: cup; female builds; 2 broods per year

Eggs: 3–4; creamy white with brown markings

Incubation: 12–13 days; female incubates

Fledging: 10–12 days; male and female feed the young

Migration: non-migrator to partial; moves around to find food

Food: insects, seeds, fruit; visits ground feeders

Compare: The American Robin (p. 227) is larger, has a red breast and lacks the white belly. The female Rose-breasted Grosbeak (p. 131) has a heavily streaked breast and obvious white eyebrows.

Stan's Notes: Named for its distinctive "tow-hee" call, given by both sexes, but known mostly for its other characteristic call, which sounds like "drink-your-tea!" Will hop backward with both feet (bilateral scratching), raking up leaf litter to locate insects and seeds. The female broods, but male does the most feeding of young. White-eyed form in southern states, red-eyed elsewhere.

male
p. 51

female

Rose-breasted Grosbeak
Pheucticus ludovicianus

Size: 7–8" (18–20 cm)

Female: Plump and heavily streaked. Large, obvious white eyebrows. Large ivory bill. Orange-to-yellow wing linings.

Male: black and white with a triangular rose patch in the center of the chest; rose wing linings

Juvenile: similar to female

Nest: cup; female and male construct; 1–2 broods per year

Eggs: 3–5; blue-green with brown markings

Incubation: 13–14 days; female and male incubate

Fledging: 9–12 days; female and male feed the young

Migration: complete, to Mexico, Central America and South America

Food: insects, seeds, fruit; comes to seed feeders

Compare: Looks like a large finch with bold white eyebrows and heavy streaking. The female Purple Finch (p. 115) has smaller eyebrows.

Stan's Notes: Both sexes sing, but the male sings much louder and clearer. Sings a rich, robin-like song with a chip note in the tune. "Grosbeak" refers to the thick, strong bill, which is used to crush seeds. Males arrive at the breeding grounds a few days before females. When females arrive, males become territorial and reduce their feeder visits. After fledging, the young visit feeders with the adults. The rose breast patch varies in size and shape in each male. Makes short flights from tree to tree with rapid wingbeats. Late to arrive in spring and early to leave in autumn.

female

male

Horned Lark

Eremophila alpestris

Size: 7–8" (18–20 cm)

Male: Tan to brown with black markings on the face. Black necklace and bill. Pale-yellow chin. Two tiny feather "horns" on the top of the head, sometimes hard to see. Dark tail with white outer tail feathers, seen in flight.

Female: duller than male; less noticeable "horns"

Juvenile: lacks a yellow chin and black markings; does not develop "horns" until the second year

Nest: ground; female builds; 2–3 broods per year

Eggs: 3–4; gray with brown markings

Incubation: 11–12 days; female incubates

Fledging: 9–12 days; female and male feed the young

Migration: non-migrator to partial migrator; moves around to find food

Food: seeds, insects

Compare: The Eastern Meadowlark (p. 305) has a yellow breast and belly. Look for the black markings by the eyes and the black necklace to identify the Horned Lark.

Stan's Notes: The only true lark native to North America. Larks are birds of open ground. Common in rural areas; often seen in large flocks. The population increased in North America over the past century as more land was cleared for farming. Male performs a fluttering courtship flight high in the air while singing a high-pitched song. Female performs a fluttering distraction display when the nest is disturbed. Starts breeding early in the year. Moves around in winter to find food. "Lark" comes from the Middle English *laverock*, or "a lark."

Wood Thrush
Hylocichla mustelina

Size: 8" (20 cm)

Male: Reddish-brown head, back and wings with color fading into a brown tail. A distinctive white breast, belly and sides, covered with black spots. White ring around black eyes, obvious on a black-streaked white face.

Female: same as male

Juvenile: similar to adult

Nest: cup; female builds; 1–2 broods per year

Eggs: 2–4; greenish blue without markings

Incubation: 13–14 days; female incubates

Fledging: 11–12 days; female and male feed the young

Migration: complete, to Mexico and Central America

Food: insects, fruit

Compare: American Robin (p. 227) is a similar body shape, but it has a red breast. Brown Thrasher (p. 155) is a similar rusty color, but it has a much longer rusty-red tail and bright-yellow eyes in comparison to the shorter brown tail and black eyes of Wood Thrush.

Stan's Notes: An easy thrush to identify due to the large dark spots on breast and belly. Well known for its liquid flute-like calls, heard deep within woodlots throughout Kentucky. Returns to the state in spring after spending the winter in Mexico and Central America. Returns to the same woodlands year after year. Often seen on the ground, hopping around like a robin in search of insects.

winter

breeding

SUMMER MIGRATION

Spotted Sandpiper
Actitis macularius

Size:	8" (20 cm)
Male:	Olive-brown back with black spots on a white chest and belly. White line over eyes. Long, dull-yellow legs. Long bill. Winter plumage lacks spots on the chest and belly.
Female:	same as male
Juvenile:	similar to winter plumage, with a darker bill
Nest:	ground; male builds; 2 broods per year
Eggs:	3–4; brownish with brown markings
Incubation:	20–24 days; male incubates
Fledging:	17–21 days; male feeds the young
Migration:	complete migrator, to southern states, Mexico and Central and South America
Food:	aquatic insects
Compare:	The Lesser Yellowlegs (p. 151) is larger. The Killdeer (p. 153) has 2 black neck bands. Look for the black spots on the chest and belly and the bobbing tail to help identify the breeding Spotted Sandpiper.

Stan's Notes: Seen along the shorelines of large ponds, lakes and rivers. One of the few shorebirds that will dive underwater when pursued. Able to fly straight up out of the water. Holds wings in a cup-like arc in flight, rarely lifting them above a horizontal plane. Walks as if delicately balanced. When standing, constantly bobs its tail. Gives a rapid series of "weet-weet-weet" calls when frightened and flying away. Female mates with multiple males and lays eggs in up to five nests. Male does all of the nest building, incubating and childcare without any help from the female.

male
p. 275

female

juvenile

Northern Cardinal
Cardinalis cardinalis

Size: 8–9" (20–23 cm)

Female: Buff-brown with red tinges on the crest and wings. Black mask and a large reddish bill.

Male: red with a large crest and bill and a black mask extending from the face to the throat

Juvenile: same as female but with a blackish-gray bill

Nest: cup; female builds; 2–3 broods per year

Eggs: 3–4; bluish white with brown markings

Incubation: 12–13 days; female and male incubate

Fledging: 9–10 days; female and male feed the young

Migration: non-migrator

Food: seeds, insects, fruit; comes to seed feeders

Compare: The Cedar Waxwing (p. 127) has a small dark bill. The juvenile Northern Cardinal (bottom inset) looks like the adult female but with a dark bill. Look for the reddish bill to identify the female Northern Cardinal.

Stan's Notes: A familiar backyard bird. Seen in a variety of habitats, including parks. Usually likes thick vegetation. One of the few species in which both females and males sing. Can be heard all year. Listen for its "whata-cheer-cheer-cheer" territorial call in spring. Watch for a male feeding a female during courtship. The male also feeds the young of the first brood while the female builds a second nest. Territorial in spring, fighting its own reflection in a window or other reflective surface. Non-territorial in winter, gathering in small flocks of up to 20 birds. Makes short flights from cover to cover, often landing on the ground. *Cardinalis* denotes importance, as represented by the red priestly garments of Catholic cardinals.

male
p. 31

female

Red-winged Blackbird
Agelaius phoeniceus

YEAR-ROUND

Size: 8½" (22 cm)

Female: Heavily streaked brown body. Pointed brown bill and white eyebrows.

Male: jet black with red-and-yellow shoulder patches (epaulets) and a pointed black bill

Juvenile: same as female

Nest: cup; female builds; 2–3 broods per year

Eggs: 3–4; bluish green with brown markings

Incubation: 10–12 days; female incubates

Fledging: 11–14 days; female and male feed the young

Migration: non-migrator to partial; moves around to find food

Food: seeds, insects; visits seed and suet feeders

Compare: The female Rose-breasted Grosbeak (p. 131) is plumper and has a thicker bill. The female Brown-headed Cowbird (p. 125) lacks streaks. Look for white eyebrows and heavy streaking to identify the female Red-winged.

Stan's Notes: One of the most widespread and numerous birds in Kentucky. In fall and winter, migrant and resident Red-wings gather in huge numbers (thousands) with other blackbirds to feed in agricultural fields, marshes, wetlands, and lakes and rivers. Flocks with as many as 10,000 birds have been reported. Males arrive before females and sing to defend their territory. The male repeats his call from the top of a cattail while showing off his red-and-yellow shoulder patches. The female chooses a mate and often builds her nest over shallow water in a thick stand of cattails. The male can be aggressive when defending the nest. Feeds mostly on seeds in spring and fall, and insects throughout the summer.

female

male

Common Nighthawk
Chordeiles minor

Size: 9" (23 cm)

Male: Camouflaged brown and white with a white chin. Distinctive white band across the wings and tail, seen only in flight.

Female: similar to male, with a tan chin; lacks a white tail band

Juvenile: similar to female

Nest: no nest; lays eggs on the ground, usually on rocks, or on rooftop; 1 brood per year

Eggs: 2; cream with lavender markings

Incubation: 19–20 days; female and male incubate

Fledging: 20–21 days; female and male feed the young

Migration: complete, to South America

Food: insects caught in the air

Compare: Male Whip-poor-will (p. 147) is similar in size, but much browner. The Chimney Swift (p. 91) is much smaller. Look for the white chin, obvious white band on the wings and characteristic flap-flap-flap-glide pattern to help identify the Common Nighthawk.

Stan's Notes: Usually only seen in flight at dusk or after sunset but not uncommon to see it sleeping on a branch during the day. A prolific insect eater and very noisy in flight, repeating a "peenting" call. Alternates slow wingbeats with bursts of quick wingbeats. Prefers to nest on flat rooftops with gravel. Populations are on the decline as gravel rooftops are converted to other styles. In spring, the male performs a showy mating ritual of a steep diving flight that ends with a loud popping noise. One of the first birds to migrate each fall.

in flight

juvenile

male

female

in-flight
juvenile

American Kestrel
Falco sparverius

YEAR-ROUND

Size: 9–11" (23–28 cm); up to 2' wingspan

Male: Rust-brown back and tail. White breast with dark spots. Two vertical black lines on a white face. Blue-gray wings. Wide black band with a white edge on the tip of a rusty tail.

Female: similar to male but slightly larger, with rust-brown wings and dark bands on the tail

Juvenile: same as adult of the same sex

Nest: cavity; does not build a nest; 1 brood per year

Eggs: 4–5; white with brown markings

Incubation: 29–31 days; male and female incubate

Fledging: 30–31 days; female and male feed the young

Migration: non-migrator to partial, to southern states and Central America

Food: insects, small mammals and birds, reptiles

Compare: The Peregrine Falcon (p. 239) is much larger and has a dark "hood" marking. No other small bird of prey has a rusty back and tail.

Stan's Notes: An unusual raptor because the sexes look different (dimorphic). Due to its small size, this falcon was once called a Sparrow Hawk. Hovers near roads, then dives for prey. Watch for it to pump its tail after landing on a perch. Perches nearly upright. Eats many grasshoppers. Adapts quickly to a wooden nest box. Can be extremely vocal, giving a loud series of high-pitched calls. Ability to see ultraviolet (UV) light helps it locate mice and other prey by their urine, which glows bright yellow in UV light.

Whip-poor-will
Antrostomas vociferus

Size: 10" (25 cm)

Male: Mottled brown and black. Distinctive black chin and a white U-shaped throat marking.

Female: same as male, but has a brown chin and tan throat marking

Juvenile: similar to adult of the same sex

Nest: no nest; lays eggs on the ground; 1–2 broods per year

Eggs: 2; white with brown markings

Incubation: 19–20 days; female and male incubate

Fledging: 18–20 days; female and male feed the young

Migration: complete, to southern coastal states, Mexico and Central America

Food: insects

Compare: Male Common Nighthawk (p. 143) has a distinctive white band across wings (seen in flight), which the Whip-poor-will lacks. Nighthawk is commonly seen flying, while Whip-poor-will is rarely seen flying.

Stan's Notes: A common bird in Kentucky, although rarely seen. Its repetitive nocturnal "whip-poor-will" call, usually heard only in the spring, is loved by many and hated by others. Nothing can be done to stop the nocturnal calling despite how much sleep you are losing. Generally found in woodlands, Whip-poor-wills sit parallel on a branch during the day. They don't build nests but lay eggs on the ground, selecting sites along forest edges. The male will care for its young if the female starts a second brood.

Northern Bobwhite
Colinus virginianus

YEAR-ROUND

Size: 10" (25 cm)

Male: Short, stocky and mostly brown with short gray tail. Prominent white eye stripe and white chin. Reddish-brown sides and belly, often with black lines and dots.

Female: similar to male, with buff-brown eye stripe and chin

Juvenile: smaller and duller than adults

Nest: ground; female and male construct; 1 brood per year

Eggs: 12–15; white to creamy without markings

Incubation: 23–24 days; female and male incubate

Fledging: 6–7 days; female and male feed the young

Migration: non-migrator

Food: insects, seeds, fruit; will come to ground feeders offering corn and millet

Compare: Ruffed Grouse (p. 179) is much larger and has neck ruffs and a feathered tuft on head. Mourning Dove (p. 159) is light brown and has a long pointed tail.

Stan's Notes: Prefers shrubs, orchards, hedgerows and pastures. Moves around in small flocks of 20 birds (often family members), called a covey. The covey often rests together during the night, in a tight circle with tails together and heads facing outward, to watch for predators. Males and females perform distraction displays when nests or young are threatened. Nest is a depression in the ground lined with grass. Often pulls nearby vegetation over nest to help conceal it. Male gives a rising whistle, "bob-white," heard mainly in spring and summer. Also gives a single "hoy" call year-round.

MIGRATION

Lesser Yellowlegs
Tringa flavipes

Size: 10–12" (25–30 cm)

Male: A typical sandpiper-type bird with a brown back and wings. Streaked white chest. Thin, straight black bill. Long yellow legs.

Female: same as male

Juvenile: same as adults

Nest: ground; female builds; 1 brood per year

Eggs: 3–4; yellowish with brown markings

Incubation: 22–23 days; male and female incubate

Fledging: 18–20 days; male and female lead the young to food

Migration: complete, to southern states, Mexico and Central and South America

Food: aquatic insects, tiny fish

Compare: The breeding Spotted Sandpiper (p. 137) has black spots on its chest.

Stan's Notes: Often seen in small flocks, combing shorelines and mudflats in search of food. Usually walks with its head down and tail up, ready to snatch up prey. Uses its long, straight bill to pluck insects and tiny fish out of the water. A member of the sandpiper group known as Tattlers, which scream alarm calls when taking off. Quite often moves into the water before taking flight and gives a variety of flight notes at takeoff. Nest is a simple depression atop a mound of earth. Nests in marshes in the spruce forests of central Alaska and Canada.

Killdeer
Charadrius vociferus

YEAR-ROUND
SUMMER

Size: 11" (28 cm)

Male: Upland shorebird with 2 black bands around the neck, like a necklace. Brown back and white belly. Bright reddish-orange rump, visible in flight.

Female: same as male

Juvenile: similar to adults, with a single neck band

Nest: ground; male scrapes; 2 broods per year

Eggs: 3–5; tan with brown markings

Incubation: 24–28 days; male and female incubate

Fledging: 25 days; male and female lead their young to food

Migration: non-migrator to partial; moves around to find food

Food: insects, worms, snails

Compare: The Spotted Sandpiper (p. 137) is found around water and lacks the 2 neck bands of the Killdeer.

Stan's Notes: The only shorebird that has two black neck bands. Technically classified as a shorebird but lives in dry habitats instead of the shore. Often found in vacant fields, gravel pits, driveways, wetland edges or along railroad tracks. Known to fake a broken wing to draw intruders away from the nest; once the nest is safe, the parent will take flight. Nests are just a slight depression in a dry area and are often hard to see. Hatchlings look like miniature adults walking on stilts. Soon after hatching, the young follow their parents around and peck for insects. Gives a loud and distinctive "kill-deer" call. Migrates in small flocks.

Brown Thrasher

Toxostoma rufum

YEAR-ROUND
SUMMER

Size: 11" (28 cm)

Male: Rust-red with a long tail. Heavy streaking on the breast and belly. Two white wing bars. Long, curved bill and bright-yellow eyes.

Female: same as male

Juvenile: same as adults but with grayish eyes

Nest: cup; female and male build; 2 broods per year

Eggs: 4–5; pale blue with brown markings

Incubation: 11–14 days; female and male incubate

Fledging: 10–13 days; female and male feed the young

Migration: non-migrator to partial; moves around to find food

Food: insects, fruit

Compare: American Robin (p. 227) and Gray Catbird (p. 225) are similar, but the Brown Thrasher is larger and has a streaked chest, rusty color and yellow eyes. Wood Thrush (p. 135) has a shorter brown tail and black eyes, compared to the longer rusty-red tail and yellow eyes of the Brown Thrasher.

Stan's Notes: A prodigious songster. Sings along forest edges and in suburban yards. Often found in thick shrubs, where it will sing deliberate musical phrases, repeating each twice. The male Brown Thrasher has the largest documented repertoire of all North American songbirds, with more than 1,100 types of songs. Builds nest low in dense shrubs, often in fencerows. Quickly flies or runs on the ground in and out of thick shrubs. A noisy feeder due to its habit of turning over leaves to find food.

male

female

Northern Flicker
Colaptes auratus

Size: 12" (30 cm)

Male: Brown and black with a black mustache and black necklace. Red spot on the nape of the neck. Speckled chest. Large white rump patch, seen only when flying.

Female: same as male but without a black mustache

Juvenile: same as adult of the same sex

Nest: cavity; female and male excavate; 1 brood per year

Eggs: 5–8; white without markings

Incubation: 11–14 days; female and male incubate

Fledging: 25–28 days; female and male feed the young

Migration: non-migrator; moves around to find food

Food: insects (especially ants and beetles); comes to suet feeders

Compare: The male Yellow-bellied Sapsucker (p. 53) has a red chin and forehead. The male Red-bellied Woodpecker (p. 59) has a red crown and lacks a mustache. Flickers are the only brown-backed woodpeckers in Kentucky.

Stan's Notes: This is the only woodpecker to regularly feed on the ground. Prefers ants and beetles and produces an antacid saliva that neutralizes the acidic defense of ants. The male often picks the nest site. Parents take up to 12 days to excavate the cavity. Can be attracted to your yard with a nest box stuffed with sawdust. Often reuses an old nest. Undulates deeply during flight, flashing yellow under its wings and tail and calling "wacka-wacka" loudly. Populations swell in winter with northern migrants.

Mourning Dove
Zenaida macroura

YEAR-ROUND

Size: 12" (30 cm)

Male: Smooth and fawn-colored. Gray patch on the head. Iridescent pink and greenish blue on the neck. Black spot behind and below the eyes. Black spots on the wings and tail. Pointed, wedged tail; white edges seen in flight.

Female: similar to male, but lacks the pink-and-green iridescent neck feathers

Juvenile: spotted and streaked plumage

Nest: platform; female and male build; 2 broods per year

Eggs: 2; white without markings

Incubation: 13–14 days; male incubates during the day, female incubates at night

Fledging: 12–14 days; female and male feed the young

Migration: non-migrator to partial migrator; will move around to find food

Food: seeds; will visit seed and ground feeders

Compare: Eurasian Collared-Dove (p. 231) has a black collar on its neck. Rock Pigeon (p. 233) is larger and has a wide range of color combinations.

Stan's Notes: Name comes from its mournful cooing. A ground feeder, bobbing its head as it walks. One of the few birds to drink without lifting its head, like the Rock Pigeon. The parents feed the young (squab) a regurgitated liquid called crop-milk during their first few days of life. Platform nest is so flimsy it often falls apart in a storm. During takeoff and in flight, wind rushes through the bird's wing feathers, creating a characteristic whistling sound.

SUMMER

Yellow-billed Cuckoo
Coccyzus americanus

Size: 12" (30 cm)

Male: Grayish-brown head, back, wings and tail. White chin, chest and belly. Undertail distinctively patterned with bold black-and-white spots and lines. Long downward-curved bill. Lower bill (mandible) is yellow.

Female: same as male

Juvenile: similar to adult, but undertail lacks bold black-and-white pattern, and bill lacks yellow

Nest: platform; female and male construct; 1–2 broods per year

Eggs: 2–6; light blue without markings

Incubation: 9–11 days; female and male incubate

Fledging: 7–9 days; female and male feed the young

Migration: complete, to South America

Food: insects

Compare: The large curved bill and bold black-and-white undertail markings make this bird hard to confuse with others.

Stan's Notes: Common summer resident throughout Kentucky. Found in a wide variety of habitats, but usually nests along forest edges. Often will place a flimsy stick nest in a densely covered tree fork. Unlike many other birds, the young do not hatch at the same time (asynchronously). There can be many days between the first and last to hatch. A very short egg-to-fledging time with some young leaving the nest (fledging) after only one week. The first to fledge are attended by the male, while the female cares for the rest in the nest. Declining in population in many states.

winter

breeding

Pied-billed Grebe

Podilymbus podiceps

Size: 12–14" (30–36 cm)

Male: Small and brown with a black chin and fluffy white patch beneath the tail. Black ring around a thick, chicken-like, ivory bill. Winter bill is brown and unmarked.

Female: same as male

Juvenile: paler than adults, with white spots and a gray chest, belly and bill

Nest: floating platform; female and male build; 1 brood per year

Eggs: 5–7; bluish white without markings

Incubation: 22–24 days; female and male incubate

Fledging: 45–60 days; female and male feed the young

Migration: non-migrator to partial, to southern states; moves around to find open water

Food: crayfish, aquatic insects, fish

Compare: Look for a puffy white patch under the tail and thick, chicken-like bill to help identify.

Stan's Notes: A common water bird during migration. Often seen diving for food. When disturbed, it slowly sinks like a submarine, quickly compressing its feathers, forcing the air out. Was called Hell-diver due to the length of time it can stay submerged. Able to surface far from where it went under. Well suited to life on water, with short wings, lobed toes, and legs set close to the rear of its body. Swims easily but moves awkwardly on land. Very sensitive to pollution. Builds nest on a floating mat in water. "Grebe" may originate from the Breton word *krib*, meaning "crest," referring to the crested head plumes of many grebes, especially during breeding season.

WINTER

Green-winged Teal
Anas crecca

Size: 14–15" (36–38 cm)

Male: Gray body. Chestnut head with a dark-green patch outlined with white from the eyes to the nape of the neck. Green patch on the wings (speculum), seen in flight. Butter-yellow tail.

Female: light-brown with black spots and a green speculum; small bill

Juvenile: same as female

Nest: ground; female builds; 1 brood per year

Eggs: 8–10; creamy white without markings

Incubation: 21–23 days; female incubates

Fledging: 32–34 days; female teaches the young to feed

Migration: complete, to Kentucky and southern states

Food: aquatic plants and insects

Compare: Male Wood Duck (p. 255) is more colorful than male Green-winged Teal. The female Blue-winged Teal (p. 167) is similar in size, but it has slight white at the base of its bill. Look for the chestnut head with a dark-green patch on each side to identify the male Green-winged Teal.

Stan's Notes: One of the smallest dabbling ducks. Tips forward in water to feed off the bottom of shallow ponds. This behavior makes it vulnerable to ingesting spent lead shot, which can cause death. It walks well on land and will also feed in flooded fields and woodlands. Known for its fast and agile flight. Groups wheel and spin through the air in tight formation. The green wing patches are most obvious during flight.

male

female

Blue-winged Teal
Spatula discors

Size: 15–16" (38–41 cm)

Male: Small, plain-looking brown duck with black speckles and a large, crescent-shaped white mark at the base of the bill. Gray head. Black tail with a small white patch. Blue wing patch (speculum), best seen in flight.

Female: duller than male, with only slight white at the base of the bill; lacks a crescent mark on the face and a white patch on the tail

Juvenile: same as female

Nest: ground; female builds; 1 brood per year

Eggs: 8–11; creamy white

Incubation: 23–27 days; female incubates

Fledging: 35–44 days; female feeds the young

Migration: partial migrator to complete, to southern states, Mexico and Central America

Food: aquatic plants, seeds, aquatic insects

Compare: Female Green-winged Teal (p. 165) lacks white near its bill. The female Mallard (p. 185) has an orange-and-black bill. Look for the white facial mark to identify the male Blue-winged.

Stan's Notes: One of the smallest ducks in North America. Constructs its nest some distance from the water. Female performs a distraction display to protect nest and young. Male leaves female near the end of incubation. Planting crops and cultivating to pond edges have caused a decline in population. Breeds as far north as Alaska. One of the longest-distance migrating ducks.

soaring

Broad-winged Hawk

Buteo platypterus

Size: 14–19" (36–48 cm); up to 3' wingspan

Male: Brown back and rust-red bars on the breast. 2 or 3 wide black-and-white tail bands. Short, round wings. White under the wings and black "fingertips," seen in flight.

Female: same as male but slightly larger

Juvenile: tail bands narrower and more numerous; brown-streaked chest and belly

Nest: platform; female and male build, but female finishes; 1 brood per year

Eggs: 2–3; off-white with brown markings

Incubation: 28–32 days; female incubates; male feeds the female during incubation

Fledging: 34–40 days; female and male feed the young

Migration: complete, to Central and South America

Food: small birds, small mammals, snakes, frogs, toads, large insects

Compare: Cooper's Hawk (p. 237) has a longer, thinner tail. The Sharp-shinned Hawk (p. 235) is much smaller. Look for the black-and-white tail bands to identify the Broad-winged.

Stan's Notes: A very common woodland hawk in eastern Kentucky. Seen in large groups (kettles) migrating in early fall. Spends most of its time hunting small birds, snakes and frogs in dense woods. Short wings help it navigate around trees. Often heard before it is seen. Screams a high-pitched whistle call repetitively when intruders are near the nest. Performs a sky-dance courtship with steep dives, sharp flights upward and rolling over.

soaring

Red-shouldered Hawk
Buteo lineatus

Size: 15–19" (38–48 cm); up to 3½' wingspan

Male: Reddish (cinnamon) head, shoulders, breast and belly. Wings and back are dark brown with white spots. Long tail with thin white bands and wide black bands. Obvious red wing linings, seen in flight.

Female: same as male

Juvenile: similar to adults but lacks the cinnamon color; white chest with dark spots

Nest: platform; female and male build; 1 brood per year

Eggs: 2–4; white with dark markings

Incubation: 27–29 days; female and male incubate

Fledging: 39–45 days; female and male feed the young

Migration: non-migrator to partial migrator; moves around to find food

Food: reptiles, amphibians, large insects, birds

Compare: The Red-tailed Hawk (p. 191) has a white chest. Cooper's Hawk (p. 237) has a slimmer body and longer tail. The Sharp-shinned Hawk (p. 235) is smaller and lacks the reddish head and belly of the Red-shouldered Hawk.

Stan's Notes: A common hawk of Kentucky forests that prefers to hunt at forest edges, spotting snakes, frogs, insects, occasional small birds and other prey as it perches. Often flaps with an alternating gliding pattern. Very vocal with a distinct scream. Breeds when it reaches 2–3 years. Remains in the same territory for many years. Starts constructing its nest in February. Young leave the nest by June.

female

male p. 61

Lesser Scaup
Aythya affinis

WINTER

Size: 16–17" (40–43 cm)

Female: Overall brown duck with dull-white patch at the base of a light gray bill. Yellow eyes.

Male: white and gray; the chest and head appear nearly black but the head looks purple with green highlights in direct sun; yellow eyes

Juvenile: same as female

Nest: ground; female builds; 1 brood per year

Eggs: 8–14; olive-buff without markings

Incubation: 22–28 days; female incubates

Fledging: 45–50 days; female teaches the young to feed

Migration: complete, to southern states, Mexico and Central America

Food: aquatic plants and insects

Compare: Female Ring-necked Duck (p. 175) is a similar size and has a white ring around the bill. Male Blue-winged Teal (p. 167) is slightly smaller and has a crescent-shaped white mark at the base of bill. Female Wood Duck (p. 181) is larger with white around eyes.

Stan's Notes: A common diving duck in Kentucky. Often in large flocks numbering in the hundreds on lakes, ponds and sewage lagoons. Completely submerges itself to feed on the bottom of lakes (unlike dabbling ducks, which only tip forward to reach bottom). Quantity of eggs (clutch size) increases with age of female. Interesting baby-sitting arrangement in which groups of young (crèches) are tended by one to three adult females. A winter resident. Doesn't breed in Kentucky.

female

male p. 65

Ring-necked Duck

Aythya collaris

Size: 16–19" (41–48 cm)

Female: Brown with a darker-brown back and crown and lighter-brown sides. Gray face. White eye-ring with a white line behind the eye. White ring around the bill. Peaked head.

Male: black head, chest and back; gray-to-white sides; blue bill with a bold white ring and a thinner ring at the base; peaked head

Juvenile: similar to female

Nest: ground; female builds; 1 brood per year

Eggs: 8–10; olive to brown without markings

Incubation: 26–27 days; female incubates

Fledging: 49–56 days; female teaches the young to feed

Migration: complete, to southern states, West Indies, Mexico and Central America

Food: aquatic plants and insects

Compare: The female Lesser Scaup (p. 173) is similar in size. Look for the white ring around the bill to help identify the female Ring-necked Duck.

Stan's Notes: A common winter resident in Kentucky. Often seen in larger lakes, usually in small flocks or just pairs. A diving duck, watch for it to dive underwater to forage for food. Springs up off the water to take flight. Has a distinctive tall, peaked head with a sloped forehead. Flattens its crown when diving. Male gives a quick series of grating barks and grunts. Female gives high-pitched peeps. Named "Ring-necked" for its cinnamon collar, which is nearly impossible to see in the field. Also called Ring-billed Duck due to the white ring on its bill.

female

male p. 63

Hooded Merganser
Lophodytes cucullatus

YEAR-ROUND

Size: 16–19" (41–48 cm)

Female: Sleek brown-and-rust bird with a red head. Ragged "hair" on the back of the head. Long, thin, brown bill.

Male: black back, rust-brown sides, long black bill; raises crest "hood" to display a white patch

Juvenile: similar to female

Nest: cavity; female lines an old woodpecker cavity or a nest box near water; 1 brood per year

Eggs: 10–12; white without markings

Incubation: 32–33 days; female incubates

Fledging: 71 days; female feeds the young

Migration: non-migrator in Kentucky

Food: small fish, aquatic insects, crustaceans (especially crayfish)

Compare: Female Lesser Scaup (p. 173) is smaller and has a dull-white patch at the base of its bill. Look for the long thin bill and ragged "hair" feathers on the back of the head to identify the female Merganser.

Stan's Notes: A small diving duck, found in shallow ponds, sloughs, lakes and rivers. Usually in small groups. Quick, low flight across the water, with fast wingbeats. Male has a deep, rolling call. Female gives a hoarse quack. Nests in wooded areas. Female will lay some eggs in the nests of other mergansers or Wood Ducks (egg dumping), resulting in 20–25 eggs in some nests. Rarely, she shares a nest, sitting with a Wood Duck.

drumming

Ruffed Grouse
Bonasa umbellus

YEAR-ROUND

Size: 16–19" (41–48 cm); up to 2' wingspan

Male: Brown chicken-like bird with a long, squared tail. Wide black band near tip of tail. Tuft of feathers on head (crest) appears like a crown when raised. Black ruffs on the sides of neck.

Female: same as male, but has less obvious neck ruffs

Juvenile: same as female

Nest: ground; female builds; 1 brood per year

Eggs: 9–12; tan with light-brown markings

Incubation: 23–24 days; female incubates

Fledging: 10–12 days; female leads the young to food

Migration: non-migrator; moves around to find food

Food: seeds, insects, fruit, leaf buds

Compare: Much larger than the Northern Bobwhite (p. 149) and lacks Bobwhite's eye stripe. Look for the feather tuft on the head and black neck ruffs to help identify the Ruffed Grouse.

Stan's Notes: A common bird of deep woods. Associated with deciduous trees, feeding on leaf buds. In the colder northern climates, scaly bristles grow on its feet during winter and serve as snowshoes. When there is enough snow, it dives into a snowbank to roost at night. In spring, the male attracts females by raising its feather tuft, fanning its tail like a turkey and standing on a log, drumming with its wings. The drumming sound is not made by its wings pounding against its chest or hitting the log, but by the air being moved by its cupped wings. Female performs a distraction display to protect her young. Two color morphs, red and gray, most apparent in the tail. Named for the black ruffs on its neck.

male
p. 255

female

Wood Duck
Aix sponsa

YEAR-ROUND

Size: 17–20" (43–51 cm)

Female: Small brown dabbling duck. Bright-white eye-ring and a not-so-obvious crest. Blue patch on wings (speculum), often hidden.

Male: highly ornamented, with a mostly green head and crest patterned with black and white; rusty chest, white belly and red eyes

Juvenile: similar to female

Nest: cavity; female lines an old woodpecker cavity or a nest box in a tree; 1 brood per year

Eggs: 10–15; creamy white without markings

Incubation: 28–36 days; female incubates

Fledging: 56–68 days; female teaches the young to feed

Migration: non-migrator to partial migrator; moves around to find open water

Food: aquatic insects, plants, seeds

Compare: The female Mallard (p. 185) and female Blue-winged Teal (p. 167) lack the eye-ring and crest. The female Northern Shoveler (p. 187) has a large, spoon-shaped bill.

Stan's Notes: A common duck of quiet, shallow backwater ponds. Nearly went extinct around 1900 due to overhunting, but it's doing well now. Nests in a tree cavity or a nest box in a tree. Seen flying in forests or perching on high branches. Female takes off with a loud, squealing call and enters the nest cavity from full flight. Lays some eggs in a neighboring nest (egg dumping), resulting in more than 20 eggs in some clutches. Hatchlings stay in the nest for 24 hours, then jump from as high as 60 feet (18 m) to the ground or water to follow their mother. They never return to the nest.

male
p. 243

female

Gadwall
Mareca strepera

Size: 19" (48 cm)

Female: Mottled brown with a pronounced color change from dark-brown body to light-brown neck and head. Bright-white wing linings, seen in flight. Small white wing patch, seen when swimming. Gray bill with orange sides.

Male: plump gray duck with a brown head and distinctive black rump, white belly, bright-white wing linings, small white wing patch, chestnut-tinged wings, gray bill

Juvenile: similar to female

Nest: ground; female lines the nest with fine grass and down feathers plucked from her chest; 1 brood per year

Eggs: 8–11; white without markings

Incubation: 24–27 days; female incubates

Fledging: 48–56 days; young feed themselves

Migration: complete, to Kentucky and southern states

Food: aquatic insects

Compare: Female Mallard (p. 185) is similar but has a blue-and-white wing mark. Look for Gadwall's white wing patch and gray bill with orange sides.

Stan's Notes: A duck of shallow marshes. Consumes mostly plant material, dunking its head in water to feed rather than tipping forward, like other dabbling ducks. Walks well on land; feeds in fields and woodlands. Nests within 300 feet (90 m) of water. Often in pairs with other duck species. Establishes pair bond during winter. A winter resident, it doesn't nest in Kentucky.

183

male
p. 257

female

Mallard

Anas platyrhynchos

YEAR-ROUND

Size:	19–21" (48–53 cm)
Female:	Brown duck with a blue-and-white wing mark (speculum). Orange-and-black bill.
Male:	large green head, white necklace, rust-brown or chestnut chest, combination of gray-and-white sides, yellow bill, orange legs and feet
Juvenile:	same as female but with a yellow bill
Nest:	ground; female builds; 1 brood per year
Eggs:	7–10; greenish to whitish, unmarked
Incubation:	26–30 days; female incubates
Fledging:	42–52 days; female leads the young to food
Migration:	non-migrator to partial migrator
Food:	seeds, plants, aquatic insects; will come to ground feeders offering corn
Compare:	Female Northern Shoveler (p. 187) is smaller and has a large, spoon-shaped bill. Female Gadwall (p. 183) has a gray bill with orange sides. The female Wood Duck (p. 181) has a white eye-ring. The female Blue-winged Teal (p. 167) is smaller than the female Mallard.

Stan's Notes: A familiar dabbling duck of lakes and ponds. Also found in rivers, streams and some backyards. Tips forward to feed on vegetation on the bottom of shallow water. The name "Mallard" comes from the Latin word *masculus,* meaning "male," referring to the male's habit of taking no part in raising the young. Female and male have white underwings and white tails, but only the male has black central tail feathers that curl upward. The female gives a classic quack. Returns to its birthplace each year.

male
p. 259

female

MIGRATION
WINTER

Northern Shoveler
Anas clypeata

Size: 19–21" (48–53 cm)

Female: Medium-size brown duck speckled with black. Green patch on the wings (speculum). Extraordinarily large, spoon-shaped bill.

Male: iridescent green head, rusty sides, white chest, large spoon-shaped bill

Juvenile: same as female

Nest: ground; female builds; 1 brood per year

Eggs: 9–12; olive without markings

Incubation: 22–25 days; female incubates

Fledging: 30–60 days; female leads the young to food

Migration: complete, to Kentucky, southern states, Mexico and Central America

Food: aquatic insects, plants

Compare: The female Wood Duck (p. 181) is smaller and has a white eye-ring. The female Mallard (p. 185) lacks the large bill. Look for the spoon-shaped bill to identify the Shoveler.

Stan's Notes: Called "Shoveler" due to the peculiar shovel-like shape of its bill. Given the common name "Northern" because it is the only species of these ducks in North America. Seen in shallow wetlands, ponds and small lakes in flocks of 5–10 birds. Flocks fly in tight formation. Swims low in water, pointing its large bill toward the water as if it's too heavy to lift. Usually swims in tight circles while feeding. Feeds mainly by filtering tiny aquatic insects and plants from the surface of the water with its bill. Female gathers plant material and forms it into a nest a short distance from the water.

female

male
p. 241

soaring

WINTER

Northern Harrier
Circus hudsonius

Size: 18–22" (45–56 cm); up to 4' wingspan

Female: Slender, low-flying hawk with a dark-brown back and brown streaking on the chest and belly. Large white rump patch. Thin black tail bands and black wing tips. Yellow eyes.

Male: silver-gray with a large white rump patch and white belly; faint, thin bands across tail; black wing tips; yellow eyes

Juvenile: similar to female, with an orange breast

Nest: ground; female and male construct; 1 brood per year

Eggs: 4–8; bluish white without markings

Incubation: 31–32 days; female incubates

Fledging: 30–35 days; male and female feed the young

Migration: complete, to Kentucky, southern states, Mexico and Central America

Food: mice, snakes, insects, small birds

Compare: Slimmer than the Red-tailed Hawk (p. 191). Look for the characteristic low gliding and the black tail bands to identify the female Harrier.

Stan's Notes: One of the easiest of hawks to identify. Glides just above the ground, following the contours of the land while searching for prey. Holds its wings just above horizontal, tilting back and forth in the wind, similar to the Turkey Vulture. Formerly called Marsh Hawk due to its habit of hunting over marshes. Feeds and nests on the ground. Will also preen and rest on the ground. Unlike other hawks, mainly uses its hearing to find prey, followed by its sight. At any age, it has a distinctive owl-like face disk.

soaring

juvenile soaring

juvenile

Red-tailed Hawk

Buteo jamaicensis

YEAR-ROUND

Size: 19–23" (48–63 cm); up to 4½' wingspan

Male: Variety of colorations, from chocolate brown to nearly all white. Often brown with a white breast and brown belly band. Rust-red tail. Underside of wing is white with a small dark patch on the leading edge near the shoulder.

Female: same as male but slightly larger

Juvenile: similar to adults, with a speckled breast and light eyes; lacks a red tail

Nest: platform; male and female build; 1 brood per year

Eggs: 2–3; white without markings or sometimes marked with brown

Incubation: 30–35 days; female and male incubate

Fledging: 45–46 days; male and female feed the young

Migration: non-migrator to partial migrator; moves around to find food

Food: small and medium-size animals, large birds, snakes, fish, insects, bats, carrion

Compare: Red-shouldered Hawk (p. 171) and Sharp-shinned Hawk (p. 235) are much smaller.

Stan's Notes: Common in open country and cities. Seen perching on fences, freeway lampposts and trees. Look for it circling above open fields and roadsides, searching for prey. Gives a high-pitched scream that trails off. Often builds a large stick nest in large trees along roads. Lines nest with finer material, like evergreen needles. Returns to the same nest site each year. The red tail develops in the second year and is best seen from above.

Barred Owl
Strix varia

YEAR-ROUND

Size: 20–24" (51–61 cm); up to 3½' wingspan

Male: Chunky brown-and-gray owl. Dark horizontal barring on upper chest. Vertical streaks on lower chest and belly. A large head and dark-brown eyes. Yellow bill and feet.

Female: same as male but slightly larger

Juvenile: light gray with a black face

Nest: cavity; does not add any nesting material; 1 brood per year

Eggs: 2–3; white without markings

Incubation: 28–33 days; female incubates

Fledging: 42–44 days; female and male feed the young

Migration: non-migrator

Food: mice, rabbits and other mammals; small birds; fish; reptiles; amphibians

Compare: Great Horned Owl (p. 195) has "horns," and the much smaller Eastern Screech-Owl (p. 223) has ears, both of which Barred Owl lacks. Look for a stocky owl with a large head and dark-brown eyes to identify the Barred Owl.

Stan's Notes: A very common owl in the state. Prefers deciduous, dense woodlands with sparse undergrowth, but it can be attracted to your yard with a simple nest box that has a large entrance hole. Often seen hunting during the day. Perches and watches for mice, birds and other prey. Hovers over water and reaches down to grab fish. After fledging, the young stay with their parents for up to four months. Often sounds like a dog barking just before calling six to eight hoots, sounding like "who-who-who-cooks-for-you."

Great Horned Owl
Bubo virginianus

Size:	21–25" (53–64 cm); up to 4' wingspan
Male:	Robust brown "horned" owl. Bright-yellow eyes and a V-shaped white throat resembling a necklace. Horizontal barring on the chest.
Female:	same as male but slightly larger
Juvenile:	similar to adults but lacks ear tufts
Nest:	no nest; takes over the nest of a crow, hawk or Great Blue Heron or uses a partial cavity, stump or broken tree; 1 brood per year
Eggs:	2–3; white without markings
Incubation:	26–30 days; female incubates
Fledging:	30–35 days; male and female feed the young
Migration:	non-migrator
Food:	mammals, birds (ducks), snakes, insects
Compare:	The Barred Owl (p. 193) has dark eyes and no "horns." The Eastern Screech-Owl (p. 223) is extremely tiny. Look for bright-yellow eyes and feather "horns" on the head to help identify the Great Horned Owl.

Stan's Notes: The largest owl in the state. The earliest nesting bird in Kentucky, it lays eggs in January and February. Able to hunt in complete darkness due to its excellent hearing. The "horns," or "ears," are tufts of feathers and have nothing to do with hearing. Cannot turn its head all the way around. Wing feathers are ragged on the ends, resulting in silent flight. Eyelids close from the top down, like ours. Fearless, it is one of the few animals that will kill skunks and porcupines. Given that, it is also called the Flying Tiger. Call sounds like "hoo-hoo-hoo-hoooo."

displaying male

non-displaying

female

Wild Turkey
Meleagris gallopavo

YEAR-ROUND

Size: 36–48" (91–122 cm)

Male: Large brown-and-bronze bird with a naked blue-and-red head. Long, straight, black beard in the center of the chest. Tail spreads open like a fan. Spurs on legs.

Female: thinner and less striking than the male; often lacks a breast beard

Juvenile: same as adult of the same sex

Nest: ground; female builds; 1 brood per year

Eggs: 10–12; buff-white with dull-brown markings

Incubation: 27–28 days; female incubates

Fledging: 6–10 days; female leads the young to food

Migration: non-migrator; moves around to find food

Food: insects, seeds, fruit

Compare: This bird is quite distinctive and unlikely to be confused with others.

Stan's Notes: The largest game bird in Kentucky, and the species from which the domestic turkey was bred. A strong flier that can approach 60 mph (97 kph). Can fly straight up, then away. Eyesight is three times better than ours. Hearing is also excellent; can hear competing males up to a mile away. Male has a "harem" of up to 20 females. Female scrapes out a shallow depression for nesting and pads it with soft leaves. Males are known as toms, females are hens, and young are poults. Roosts in trees at night. Eliminated from many states due to market hunting and loss of habitat, and reintroduced widely from the 1960s to the 1980s. Populations are now stable and growing.

WINTER

Ruby-crowned Kinglet
Regulus calendula

Size: 4" (10 cm)

Male: Small, teardrop-shaped green-to-gray bird. Two white wing bars and a white eye-ring. Hidden ruby crown.

Female: same as male, but lacks a ruby crown

Juvenile: same as female

Nest: pendulous; female builds; 1 brood per year

Eggs: 4–5; white with brown markings

Incubation: 11–12 days; female incubates

Fledging: 11–12 days; female and male feed the young

Migration: complete, to Kentucky, southern states and Mexico

Food: insects, berries

Compare: The Golden-crowned Kinglet (p. 201) is similar but lacks the ruby crown. The female American Goldfinch (p. 295) shares the drab olive plumage and unmarked chest, but it is larger. Look for the white eye-ring to identify the Ruby-crowned Kinglet.

Stan's Notes: The third-smallest bird in the state. Look for it flitting around thick shrubs low to the ground. It takes a quick eye to see the ruby crown, which the male flashes when he is excited. The female weaves an unusually intricate nest and fastens colorful lichens and mosses to the exterior with spiderwebs. Sings a distinctive song that starts out soft and ends loud and on a higher note. "Kinglet" originates from the word *king,* referring to the male's red crown, and the diminutive suffix *let,* meaning "small."

male

female

WINTER

Golden-crowned Kinglet
Regulus satrapa

Size: 4" (10 cm)

Male: Tiny, plump green-to-gray bird. Distinctive yellow-and-orange patch with a black border on the crown (top inset). White eyebrow mark. 2 white wing bars.

Female: same as male, but has a yellow crown with a black border, lacks any orange (bottom inset)

Juvenile: same as adults, but lacks gold on crown

Nest: pendulous; female builds; 1–2 broods per year

Eggs: 5–9; white or creamy with brown markings

Incubation: 14–15 days; female incubates

Fledging: 14–19 days; female and male feed the young

Migration: complete, to Kentucky and southern states

Food: insects, fruit, tree sap

Compare: Similar to the Ruby-crowned Kinglet (p. 199), but the Golden-crowned has an obvious crown. Female American Goldfinch (p. 295) is larger and has an all-black forehead.

Stan's Notes: Common during winter in Kentucky, but might be more frequently seen during migration when flocks are moving through the state. Often in flocks with chickadees, nuthatches, woodpeckers, Brown Creepers and Ruby-crowned Kinglets. Flicks its wings when moving around. Can have so many eggs in its small nest that eggs are in two layers. Drinks tree sap and feeds by gleaning insects from trees. Can be very tame and approachable.

Red-breasted Nuthatch

Sitta canadensis

YEAR-ROUND WINTER

Size: 4½" (11 cm)

Male: Gray-backed bird with an obvious black eye line and black cap. Rust-red breast and belly.

Female: duller than male and has a gray cap and pale undersides

Juvenile: same as female

Nest: cavity; male and female excavate a cavity or move into a vacant hole; 1 brood per year

Eggs: 5–6; white with red brown markings

Incubation: 11–12 days; female incubates

Fledging: 14–20 days; female and male feed the young

Migration: irruptive; moves around the state during winter in search of food

Food: insects, insect eggs, seeds; comes to seed and suet feeders

Compare: The White-breasted Nuthatch (p. 207) is larger and has a white breast. Look for the rust-red breast and black eye line to help identify the Red-breasted Nuthatch.

Stan's Notes: The nuthatch climbs down trunks of trees headfirst, searching for insects. Like a chickadee, it grabs a seed from a feeder and flies off to crack it open. Wedges the seed into a crevice and pounds it open with several sharp blows. The name "Nuthatch" comes from the Middle English moniker *nuthak*, referring to the habit of hacking seeds open. Look for it in mature conifers, where it extracts seeds from pine cones. Excavates a cavity or takes an old woodpecker hole or a natural cavity and builds a nest within. Gives a series of nasal "yank-yank-yank" calls.

Carolina Chickadee
Poecile carolinensis

YEAR-ROUND

Size: 5" (13 cm)

Male: Mostly gray with a black cap and chin. White face and chest with a tan belly. Darker-gray tail.

Female: same as male

Juvenile: same as adult

Nest: cavity; female and male build or excavate; 1–2 broods per year

Eggs: 5–7; white with reddish-brown markings

Incubation: 11–12 days; female and male incubate

Fledging: 13–17 days; female and male feed the young

Migration: non-migrator

Food: insects, seeds, fruit; comes to seed and suet feeders

Compare: Tufted Titmouse (p. 215) is a close relative, but it has an erect crest and lacks the black cap and chin.

Stan's Notes: Widespread bird in Kentucky. One of the first birds to use a newly placed feeder. Flies to a feeder, grabs a seed and carries it to a branch. To get to the meat inside, it holds the seed down with its feet and hammers the shell open with its bill. Returns for another seed. A friendly bird. Can be tamed and hand-fed. Attracted with a nest box that has a 1¼-inch entrance hole. Female gives a loud snake-like hiss if disturbed on the nest. Often seen with other birds (mixed flock) in winter. Song is a high, fast "chika-dee-dee-dee-dee."

White-breasted Nuthatch
Sitta carolinensis

Size: 5–6" (13–15 cm)

Male: Slate gray with a white face, breast and belly. Large white patch on the rump. Black cap and nape. Bill is long and thin, slightly upturned. Chestnut undertail.

Female: similar to male, but has a gray cap and nape

Juvenile: similar to female

Nest: cavity; female and male build a nest within; 1 brood per year

Eggs: 5–7; white with brown markings

Incubation: 11–12 days; female incubates

Fledging: 13–14 days; female and male feed the young

Migration: non-migrator

Food: insects, insect eggs, seeds; comes to seed and suet feeders

Compare: The Red-breasted Nuthatch (p. 203) is smaller and has a rust-red belly and distinctive black eye line. Look for the white breast to help identify the White-breasted Nuthatch.

Stan's Notes: The nuthatch hops headfirst down trees, looking for insects missed by birds climbing up. Its climbing agility is due to an extra-long hind toe claw, or nail, that is nearly twice the size of its front claws. "Nuthatch," from the Middle English *nuthak*, refers to the bird's habit of wedging a seed in a crevice and hacking it open. Often seen in flocks with chickadees, Brown Creepers and Downy Woodpeckers. Mates stay together year-round, defending a small territory. Gives a characteristic "whi-whi-whi-whi" spring call during February and March.

207

male

female

first winter

Yellow-rumped Warbler
Setophaga coronata

Size: 5–6" (13–15 cm)

Male: Slate gray with black streaking on the chest. Yellow patches on the head, flanks and rump. White chin and belly. Two white wing bars.

Female: duller gray than the male, mixed with brown

Juvenile: first winter is similar to the adult female

Nest: cup; female builds; 2 broods per year

Eggs: 4–5; white with brown markings

Incubation: 12–13 days; female incubates

Fledging: 10–12 days; female and male feed young

Migration: complete, to Kentucky, southern states, Mexico and Central America

Food: insects, berries; visits suet feeders in spring

Compare: Magnolia Warbler (p. 287) is more yellow than Yellow-rumped. The Prairie Warbler (p. 289) has an olive back with chestnut streaks. Male Common Yellowthroat (p. 293) has a yellow breast and a very distinctive black mask. Look for a combination of yellow patches on the rump, flanks and head of Yellow-rumped Warbler.

Stan's Notes: One of the most common warblers, seen in flocks of hundreds during spring and fall migrations. One of the few warblers to spend the winter in Kentucky. Familiar call is a single robust "chip," heard mostly during migration and winter. Sings a wonderful song in spring. In the fall, the male molts to a dull color similar to the female, but he retains his yellow patches all year. Also called Myrtle Warbler. Sometimes called Butter-butt due to the yellow patch on its rump.

Yellow-throated Warbler
Setophaga dominica

Size: 5½" (14 cm)

Male: A gray-backed warbler with a bright-yellow throat. Black streaks on white belly. A white spot on neck and white eyebrows.

Female: same as male, but duller with browner back

Juvenile: similar to female

Nest: cup; female and male construct; 1–2 broods per year

Eggs: 4–5; gray with dark markings

Incubation: 12–13 days; female incubates

Fledging: 10–12 days; female and male feed young

Migration: complete, to southern states, Mexico and Central America

Food: insects

Compare: Similar to Magnolia Warbler (p. 287), but Magnolia has a yellow belly with black streaks, unlike Yellow-throated's white belly with black streaks. The Yellow-rumped Warbler (p. 209) has yellow on the rump, flanks and head and lacks a yellow throat. The Prairie Warbler (p. 289) is slightly smaller and lacks Yellow-throated's gray back.

Stan's Notes: One of the most widespread of Kentucky's nesting warblers. Among the first returning warblers to the state, usually around mid-April. Prefers cypress and oak woodlands. Finds food by creeping along and searching underneath vertical surfaces such as tree bark. Highly attracted to water, it is often seen bathing in any puddle of water. In some, the white eyebrows are tinged yellow. It is rarely a Brown-headed Cowbird host.

female
p. 109

male

Dark-eyed Junco

Junco hyemalis

Size: 5½" (14 cm)

Male: Plump, dark-eyed bird with a slate-gray-to-charcoal chest, head and back. White belly. Pink bill. White outer tail feathers appear like a white V in flight.

Female: round with brown plumage

Juvenile: similar to female, with streaking on the breast and head

Nest: cup; female and male build; 2 broods per year

Eggs: 3–5; white with reddish-brown markings

Incubation: 12–13 days; female incubates

Fledging: 10–13 days; male and female feed the young

Migration: complete, throughout the U.S.; winters in Kentucky; non-migrator in parts of eastern Kentucky

Food: seeds, insects; visits ground and seed feeders

Compare: Rarely confused with any other bird. Look for the pink bill and small flocks feeding under feeders to identify the male Dark-eyed Junco.

Stan's Notes: A common winter bird of Kentucky. Winter flocks migrate from Canada to Kentucky and beyond. Adheres to a rigid social hierarchy, with dominant birds chasing the less dominant ones. Look for the white outer tail feathers flashing in flight. Often seen in small flocks on the ground, where it uses its feet to simultaneously "double-scratch" to expose seeds and insects. Eats many weed seeds. Nests in a wide variety of wooded habitats during April and May. Several subspecies of Dark-eyed Junco were previously considered to be separate species. Males don't go as far south as females in winter.

Tufted Titmouse
Baeolophus bicolor

Size:	6" (15 cm)
Male:	Slate gray with a white chest and belly. Pointed crest. Rust-brown wash on the flanks. Gray legs and dark eyes.
Female:	same as male
Juvenile:	same as adult
Nest:	cavity; female lines an old woodpecker cavity; 2 broods per year
Eggs:	5–7; white with brown markings
Incubation:	13–14 days; female incubates
Fledging:	15–18 days; female and male feed the young
Migration:	non-migrator
Food:	insects, seeds, fruit; will come to seed and suet feeders
Compare:	The Carolina Chickadee (p. 205) is a close relative but is smaller and lacks a crest. The White-breasted Nuthatch (p. 207) has a rust-brown undertail. Look for the pointed crest to help identify the Tufted Titmouse.

Stan's Notes: A common feeder bird that can be attracted with an offering of black oil sunflower seeds or suet. Can also be attracted with a nest box. Well known for its "peter-peter-peter" call, which it quickly repeats. Notorious for pulling hair from sleeping dogs, cats and squirrels to line its nest. Usually seen only one or two at a time. Male feeds female during courtship and nesting. The prefix *tit* in the common name comes from a Scandinavian word meaning "little." Suffix *mouse* is derived from the Old English word *mase*, meaning "bird." Simply translated, it is a "small bird."

Eastern Phoebe
Sayornis phoebe

Size: 7" (18 cm)

Male: Plain gray with slightly darker wings, a light-olive belly and a thin, dark bill.

Female: same as male

Juvenile: same as adults

Nest: cup; female builds; 2 broods per year

Eggs: 4–5; white without markings

Incubation: 15–16 days; female incubates

Fledging: 15–16 days; male and female feed the young

Migration: complete, to southern states and Mexico

Food: insects

Compare: The Gray Catbird (p. 225) has a black crown and a chestnut patch under its tail. Eastern Phoebe lacks any distinctive markings. Listen for its well-enunciated "fee-bee" call and look for the hawking and tail-pumping behaviors to help identify this bird.

Stan's Notes: A sparrow-size bird that often perches on the end of a dead branch. Found in forests, yards and farms. In a process called hawking, it waits for a passing insect. When a bug flies near, it launches out to catch it and then returns to the same branch. It has a distinctive habit of pumping its tail up and down while perching. Builds nest beneath the eaves of houses, under bridges or in other sheltered spots. Uses mud, grass and moss for nest materials and hair (and sometimes feathers) for the lining. The common name is derived from its distinct "fee-bee" call, which it repeats over and over from the top of dead branches.

Eastern Kingbird

Tyrannus tyrannus

SUMMER

Size: 8" (20 cm)

Male: Mostly gray and black with a white chin and belly. Black head and tail with a distinct white band on the tip of the tail. Concealed red crown, rarely seen.

Female: same as male

Juvenile: same as adults

Nest: cup; male and female build; 1 brood per year

Eggs: 3–4; white with brown markings

Incubation: 16–18 days; female incubates

Fledging: 16–18 days; female and male feed the young

Migration: complete, to South America

Food: insects, fruit

Compare: The American Robin (p. 227) is larger and has a rust-red breast. The Eastern Phoebe (p. 217) is smaller and has an olive-green belly. Look for the white tail band to identify the Kingbird.

Stan's Notes: A summer resident seen in open fields and prairies. Autumn migration begins in late August and early September, with as many as 20 birds in a group. Returns to the mating ground in spring, where pairs defend their territory. Seems to be unafraid of other birds and chases larger birds. Given the common name "King" for its bold attitude and behavior. In a hunting technique known as hawking, it perches on a branch and watches for insects, flies out to catch one, and then returns to the same perch. Swoops from perch to perch when hunting. Becomes very vocal during late summer, when family members call back and forth to one another while hunting for insects.

SUMMER

Great Crested Flycatcher
Myiarchus crinitus

Size: 8" (20 cm)

Male: Gray head with a prominent crest. Gray back and throat. Yellow from the belly to the base of a reddish-brown tail. Lower bill is yellow at the base.

Female: same as male

Juvenile: same as adults

Nest: cavity; female and male construct; 1 brood per year

Eggs: 4–6; white or buff with brown markings

Incubation: 13–15 days; female incubates

Fledging: 14–21 days; female and male feed the young

Migration: complete, to Mexico and Central and South America

Food: insects, fruit

Compare: The Eastern Kingbird (p. 219) has a white band across its tail. The Eastern Phoebe (p. 217) is similar, but it lacks a crest and yellow belly. Look for the crest to identify the Flycatcher.

Stan's Notes: Breeds mostly in western Kentucky. It lives high up in trees, rarely coming to the ground. Makes long flights from tree-top to treetop, moving from one hunting area to another. Gleans insects from tree leaves. Often heard before seen. "Great Crested" refers to the set of extra-long feathers on top of its head (crest), which the bird raises when alert or agitated, like the Northern Cardinal. Nests in an old woodpecker hole but can be attracted with a man-made nest box that has an entrance hole 1½–2½ inches (4–6 cm) in diameter. Often stuffs the cavity with a collection of fur, feathers, string and snake skins.

red morph

gray morph

Eastern Screech-Owl
Megascops asio

Size: 8–10" (20–25 cm); up to 2' wingspan

Male: Small "eared" owl that occurs in different colorations. Gray morph is mottled gray and white. Red morph is mottled rust and white. Short wings. Bright-yellow eyes.

Female: same as male but slightly larger

Juvenile: lighter color than adults of the same morph and usually lacks ear tufts

Nest: cavity, old woodpecker cavity or man-made nest box; does not add any nesting material; 1 brood per year

Eggs: 4–5; white without markings

Incubation: 25–26 days; female incubates, male feeds the female during incubation

Fledging: 26–27 days; male and female feed the young

Migration: non-migrator; moves around in winter

Food: large insects, small mammals, birds, snakes

Compare: This is the only small owl in Kentucky with ear tufts. Can be gray or rust in color.

Stan's Notes: Commonly found in forests that have suitable natural cavities for nesting and roosting. Active from dusk to dawn. Usually gives a tremulous, descending trill, like a sound effect in a scary movie. Seldom gives a screeching call. Often seen sunning itself at a nest-box hole during winter. Mates may have a long-term pair bond and may roost together at night. Excellent hearing and eyesight. Flaps rapidly and flies silently. Has winter and summer territories. The gray morph is more common than the red.

SUMMER

Gray Catbird
Dumetella carolinensis

Size: 9" (23 cm)

Male: Handsome slate-gray bird with a black crown and a long, thin, black bill. Often lifts up its tail, exposing a chestnut patch beneath.

Female: same as male

Juvenile: same as adults

Nest: cup; female and male build; 2 broods per year

Eggs: 4–6; blue-green without markings

Incubation: 12–13 days; female incubates

Fledging: 10–11 days; female and male feed young

Migration: complete, to southern coastal states, Mexico and Central America

Food: insects, occasional fruit; visits suet feeders

Compare: The Eastern Phoebe (p. 217) is smaller and has an olive belly. The Eastern Kingbird (p. 219) is similar in size but has a white belly and a white band across its tail. To identify the Gray Catbird, look for the black crown and chestnut patch under the tail.

Stan's Notes: A secretive bird, more often heard than seen. The Chippewa Indians gave it a name that means "the bird that cries with grief" due to its raspy call. Called "Catbird" because the sound is like the meowing of a house cat. Often mimics other birds, rarely repeating the same phrases. Builds its nest with small twigs. Nests in thick shrubs and quickly flies back into shrubs if approached. If a cowbird lays an egg in its nest, the catbird will quickly break it and eject it.

American Robin
Turdus migratorius

YEAR-ROUND

Size: 9–11" (23–28 cm)

Male: Familiar gray bird with a dark rust-red breast and a nearly black head and tail. White chin with black streaks. White eye-ring.

Female: similar to male, with a duller rust-red breast and a gray head

Juvenile: similar to female, with a speckled breast and brown back

Nest: cup; female builds with help from the male; 2–3 broods per year

Eggs: 4–7; pale blue without markings

Incubation: 12–14 days; female incubates

Fledging: 14–16 days; female and male feed the young

Migration: non-migrator to partial; moves around to find food

Food: insects, fruit, berries, earthworms

Compare: Familiar bird to all. To differentiate the male from the female, compare the nearly black head and rust-red chest of the male with the gray head and duller chest of the female.

Stan's Notes: Although they can be seen year-round, many of Kentucky's robins will migrate to southern states. Most return by late February and are nesting by March. Can be heard singing all night long in spring. A robin isn't listening for worms when it turns its head to one side. It is focusing its sight out of one eye to look for dirt moving, which is caused by worms moving. Territorial, often fighting its reflection in a window. Males have dark heads and a brighter red breast than females.

displaying

Northern Mockingbird
Mimus polyglottos

Size: 10" (25 cm)

Male: Silvery-gray head and back with a light-gray breast and belly. White wing patches, seen in flight or during display. Tail mostly black with white outer tail feathers. Black bill.

Female: same as male

Juvenile: dull gray with a heavily streaked breast and a gray bill

Nest: cup; female and male construct; 2 broods per year, sometimes more

Eggs: 3–5; blue-green with brown markings

Incubation: 12–13 days; female incubates

Fledging: 11–13 days; female and male feed the young

Migration: non-migrator to partial migrator; moves around to find food

Food: insects, fruit

Compare: The Gray Catbird (p. 225) is slate gray and lacks wing patches. Look for Mockingbird to spread its wings, flash its white wing patches and wag its tail from side to side.

Stan's Notes: A very animated bird. Performs an elaborate mating dance. Facing each other with heads and tails erect, pairs will run toward each other, flashing their white wing patches, and then retreat to cover nearby. Thought to flash the wing patches to scare up insects when hunting. Sits for long periods on top of shrubs. Imitates other birds (vocal mimicry); hence the common name. Young males often sing at night. Often unafraid of people, allowing for close observation.

Eurasian Collared-Dove
Streptopelia decaocto

YEAR-ROUND

Size: 12½" (32 cm)

Male: Head, neck, breast and belly are gray to tan. Back, wings and tail are slightly darker. Thin black collar with a white border on the nape of the neck. Tail is long and squared.

Female: same as male

Juvenile: similar to adults

Nest: platform; female and male build; 2–3 broods per year

Eggs: 3–5; creamy white without markings

Incubation: 12–14 days; female and male incubate

Fledging: 12–14 days; female and male feed the young

Migration: non-migrator

Food: seeds; will visit ground and seed feeders

Compare: The Mourning Dove (p. 159) is slightly smaller and darker. The Rock Pigeon (p. 233) has colorful iridescent patches. Look for the black collar on the nape and the squared tail to help identify the Eurasian Collared-Dove.

Stan's Notes: A non-native bird. Moved into Florida in the 1980s after inadvertent introduction to the Bahamas. It has been expanding its range across North America and is predicted to spread just like it did through Europe from Asia. Unknown how this "new" bird will affect populations of the native Mourning Dove. Nearly identical to the Ringed Turtle-Dove, a common pet bird. The dark mark on the back of the neck gave rise to the common name. Look for flashes of white in the tail and dark wing tips when it lands or takes off.

Rock Pigeon
Columba livia

YEAR-ROUND

Size: 13" (33 cm)

Male: No set color pattern. Shades of gray to white with patches of gleaming, iridescent green and blue. Often has a light rump patch.

Female: same as male

Juvenile: same as adults

Nest: platform; female builds; 3–4 broods per year

Eggs: 1–2; white without markings

Incubation: 18–20 days; female and male incubate

Fledging: 25–26 days; female and male feed the young

Migration: non-migrator

Food: seeds

Compare: The Eurasian Collared-Dove (p. 231) has a black collar on the nape. The Mourning Dove (p. 159) is smaller and light brown and lacks the variety of color combinations of the Rock Pigeon.

Stan's Notes: Also known as the Domestic Pigeon. Formerly known as the Rock Dove. Introduced to North America from Europe by the early settlers. Most common around cities and barnyards, where it scratches for seeds. One of the few birds with a wide variety of colors, produced by years of selective breeding while in captivity. Parents feed the young a regurgitated liquid known as crop-milk for the first few days of life. One of the few birds that can drink without tilting its head back. Nests under bridges or on buildings, balconies, barns and sheds. Was once thought to be a nuisance in cities and was poisoned. Now, many cities have Peregrine Falcons (p. 239) feeding on Rock Pigeons, which keeps their numbers in check.

soaring

juvenile

Sharp-shinned Hawk

Accipiter striatus

YEAR-ROUND

Size: 10–14" (25–36 cm); up to 2' wingspan

Male: Small woodland hawk with a gray back and head and a rust-red chest. Short wings. Long, squared tail and several dark tail bands, with the widest at the end of the tail. Red eyes.

Female: same as male but larger

Juvenile: same size as adults, with a brown back, heavy streaking on the chest and yellow eyes

Nest: platform; female builds; 1 brood per year

Eggs: 4–5; white with brown markings

Incubation: 32–35 days; female incubates

Fledging: 24–27 days; female and male feed the young

Migration: non-migrator to partial; moves around in winter

Food: birds, small mammals

Compare: Cooper's Hawk (p. 237) is larger and has a larger head, a slightly longer neck and a rounded tail. Red-shouldered Hawk (p. 171) is larger and has a reddish head and belly. Look for the squared tail to help identify the Sharp-shinned Hawk.

Stan's Notes: A common hawk of backyards, parks and woodlands. Constructs its nest with sticks, usually high in a tree. Typically seen swooping in on birds at feeders and chasing them. Its short wingspan and long tail help it maneuver and pursue prey. Calls a loud, high-pitched "kik-kik-kik-kik." Named "Sharp-shinned" for the sharp projection (keel) on the leading edge on the tarsus bone of its foot. In most birds, the tarsus bone is rounded, not sharp.

soaring

juvenile

Cooper's Hawk
Accipiter cooperii

YEAR-ROUND

Size: 14–20" (36–51 cm); up to 3' wingspan

Male: Medium-size hawk with short wings and a long, rounded tail with several black bands. Slate-gray back, rusty breast, dark wing tips. Gray bill with a bright-yellow spot at the base. Dark-red eyes.

Female: similar to male but larger

Juvenile: brown back, brown streaking on the breast, bright-yellow eyes

Nest: platform; male and female construct; 1 brood per year

Eggs: 2–4; greenish with brown markings

Incubation: 32–36 days; female and male incubate

Fledging: 28–32 days; male and female feed the young

Migration: non-migrator to partial migrator; moves around in winter

Food: small birds, mammals

Compare: The Sharp-shinned Hawk (p. 235) is smaller and lighter gray and has a squared tail. Look for the banded, rounded tail to help identify Cooper's Hawk.

Stan's Notes: A common year-round resident hawk, found in many habitats, from woodlands to parks and backyards. Stubby wings help it to navigate around trees while it chases small birds. Will ambush prey, flying into heavy brush or even running on the ground in pursuit. Comes to feeders, hunting for birds. Flies with long glides followed by a few quick flaps. Calls a loud, clear "cack-cack-cack-cack." The young have gray eyes that turn bright yellow at 1 year and turn dark red later, after 3–5 years.

juvenile

in-flight juvenile

in flight

Peregrine Falcon
Falco peregrinus

YEAR-ROUND MIGRATION

Size: 16–20" (41–51 cm); up to 3¾' wingspan

Male: Dark-gray back and tan-to-white chest. Horizontal bars on belly, legs and undertail. Dark "hood" head marking and wide black mustache. Yellow base of bill and eye-ring. Yellow legs.

Female: similar to male but noticeably larger

Juvenile: overall darker than adults, with heavy streaking on the chest and belly

Nest: ground (scrape) on a cliff edge, tall building, bridge or smokestack; 1 brood per year

Eggs: 3–4; white, some with brown markings

Incubation: 29–32 days; female and male incubate

Fledging: 35–42 days; male and female feed the young

Migration: partial to non-migrator in parts of Kentucky

Food: birds (Rock Pigeons in cities, shorebirds and waterfowl in rural areas)

Compare: The American Kestrel (p. 145) is smaller and has 2 vertical black stripes on its face. Look for the dark "hood" head marking and mustache marks to identify the Peregrine Falcon.

Stan's Notes: A wide-bodied raptor that hunts many bird species. The larger females hunt larger prey. Lives in many cities, diving (stooping) on pigeons at speeds of up to 200 mph (322 kph), which knocks them to the ground. Soars with its wings flat, often riding thermals. During courtship, the male brings food to the female and performs aerial displays. Likes to nest on a high ledge or platform for a good view of its territory. A solitary nester and monogamous.

female
p. 189

male

soaring

Northern Harrier
Circus hudsonius

Size: 18–22" (45–56 cm); up to 4' wingspan

Male: Slender, low-flying hawk. Silver-gray with a large white rump patch and white belly. Long tail with faint narrow bands. Black wing tips. Yellow eyes.

Female: dark-brown back, brown streaking on breast and belly, large white rump patch, thin black tail bands, black wing tips, yellow eyes

Juvenile: similar to female, with an orange breast

Nest: ground; female and male construct; 1 brood per year

Eggs: 4–8; bluish white without markings

Incubation: 31–32 days; female incubates

Fledging: 30–35 days; male and female feed the young

Migration: complete, to Kentucky, southern states, Mexico and Central America

Food: mice, snakes, insects, small birds

Compare: Slimmer than the Red-tailed Hawk (p. 191). The Cooper's Hawk (p. 237) has a rusty breast. Look for a low-gliding hawk with a large white rump patch to identify the male Harrier.

Stan's Notes: One of the easiest of hawks to identify. Glides just above the ground, following the contours of the land while searching for prey. Holds its wings just above horizontal, tilting back and forth in the wind, similar to Turkey Vultures. Formerly called the Marsh Hawk due to its habit of hunting over marshes. Feeds and nests on the ground. Will also preen and rest on the ground. Unlike other hawks, mainly uses its hearing to find prey, followed by sight. At any age, has a distinctive owl-like face disk.

female
p. 183

male

Gadwall

Mareca strepera

WINTER

Size: 19" (48 cm)

Male: A plump gray duck with a brown head and a distinctive black rump. White belly. Chestnut-tinged wings. Bright-white wing linings. Small white wing patch, seen when swimming. Gray bill.

Female: similar to female Mallard, a mottled brown with a pronounced color change from dark-brown body to light-brown neck and head, bright-white wing linings, small white wing patch, gray bill with orange sides

Juvenile: similar to female

Nest: ground; female lines the nest with fine grass and down feathers plucked from her chest; 1 brood per year

Eggs: 8–11; white without markings

Incubation: 24–27 days; female incubates

Fledging: 48–56 days; young feed themselves

Migration: complete, to Kentucky and southern states

Food: aquatic insects

Compare: Male Gadwall is one of the few gray ducks. Look for its distinctive black rump.

Stan's Notes: A duck of shallow marshes. Consumes mostly plant material, dunking its head in water to feed rather than tipping forward, like other dabbling ducks. Walks well on land; feeds in fields and woodlands. Frequently in pairs with other duck species. Nests within 300 feet (90 m) of water. Establishes pair bond in winter. A winter resident, it doesn't nest in Kentucky.

in flight

Canada Goose
Branta canadensis

YEAR-ROUND

Size:	25–43" (64–109 cm); up to 5½' wingspan
Male:	Large gray goose with a black neck and head. White chin and cheek strap.
Female:	same as male
Juvenile:	same as adults
Nest:	platform, on the ground; female builds; 1 brood per year
Eggs:	5–10; white without markings
Incubation:	25–30 days; female incubates
Fledging:	42–55 days; male and female teach the young to feed
Migration:	non-migrator; moves around in winter to find food
Food:	aquatic plants, insects, seeds
Compare:	rarely confused with any other bird

Stan's Notes: Calls a classic "honk-honk-honk," especially in flight. Flocks fly in a large V when traveling long distances. Begins breeding in the third year. Adults mate for many years. If threatened, they will hiss as a warning. Males stand as sentinels at the edge of their group and will bob their heads and become aggressive if approached. Adults molt their primary flight feathers while raising their young, rendering family groups temporarily flightless. Several subspecies occur in the U.S. Generally eastern groups are paler than western. Their size also varies, decreasing northward. The smallest subspecies is in the Arctic.

245

in flight

YEAR-ROUND

Great Blue Heron
Ardea herodias

Size: 42–48" (107–122 cm); up to 6' wingspan

Male: Tall and gray. Black eyebrows end in long plumes at the back of the head. Long yellow bill. Long feathers at the base of the neck drop down in a kind of necklace. Long legs.

Female: same as male

Juvenile: same as adults, but more brown than gray, with a black crown; lacks plumes

Nest: platform in a colony; male and female build; 1 brood per year

Eggs: 3–5; blue-green without markings

Incubation: 27–28 days; female and male incubate

Fledging: 56–60 days; male and female feed the young

Migration: non-migrator; moves around to find food

Food: small fish, frogs, insects, snakes, baby birds

Compare: Similar size as the Sandhill Crane (p. 249), but lacks the Crane's red crown. Look for the long, yellow bill to help identify the Great Blue Heron.

Stan's Notes: One of the most common herons. Found in open water, from small ponds to large lakes. Stalks small fish in shallow water. Will strike at mice, squirrels and nearly anything it comes across. Red-winged Blackbirds will attack it to stop it from taking their babies out of the nest. In flight, it holds its neck in an S shape and slightly cups its wings, while the legs trail straight out behind. Nests in a colony of up to 100 birds. Nests in trees near or hanging over water. Barks like a dog when startled.

in flight

rusty
stain

in-flight
rusty stain

MIGRATION

Sandhill Crane
Grus canadensis

Size: 42–48" (107–122 cm); up to 7' wingspan

Male: Elegant gray crane with long legs and neck. Wings and body often rust brown from mud staining. Scarlet-red cap. Yellow to red eyes.

Female: same as male

Juvenile: dull brown with yellow eyes; lacks a red cap

Nest: ground; female and male construct; 1 brood per year

Eggs: 2; olive with brown markings

Incubation: 28–32 days; female and male incubate

Fledging: 65 days; female and male feed the young

Migration: complete, to Florida, Georgia and other southern states

Food: insects, fruit, worms, plants, amphibians

Compare: Great Blue Heron (p. 247) has a longer bill and holds its neck in an S shape during flight. Look for the scarlet-red cap to help identify the Sandhill Crane.

Stan's Notes: Preens mud into its feathers, staining its plumage rust brown (see insets). Gives a very loud and distinctive rattling call, often heard before the bird is seen. Flight is characteristic, with a faster upstroke, making the wings look like they're flicking in flight. Can fly at heights of over 10,000 feet (3,050 m). Nests on the ground in a large mound of aquatic vegetation. Performs a spectacular mating dance: The birds will face each other, then bow and jump into the air while making loud cackling sounds and flapping their wings. They will also flip sticks and grass into the air during their dance.

male

female

Ruby-throated Hummingbird
Archilochus colubris

Size: 3–3½" (7.5–9 cm)

Male: Tiny iridescent green bird with black throat patch that reflects bright ruby red in sun.

Female: same as male, but lacking the throat patch

Juvenile: same as female

Nest: cup; female builds; 1–2 broods per year

Eggs: 2; white without markings

Incubation: 12–14 days; female incubates

Fledging: 14–18 days; female feeds the young

Migration: complete, to southern states, Mexico and Central America

Food: nectar, insects; will come to nectar feeders

Compare: No other bird is as tiny. The Sphinx Moth also hovers at flowers but has clear wings, doesn't hum in flight, moves much slower than the Ruby-throated and can be approached.

Stan's Notes: This is the smallest bird in the state. Can fly straight up, straight down or backward and hover in midair. Does not sing but chatters or buzzes to communicate. Weighs about the same as a U.S. penny; it takes about five average-size hummingbirds to equal the weight of one chickadee. The wings create the humming sound. Flaps 50–60 times or more per second when flying at top speed. Breathes 250 times per minute. Heart beats up to 1,260 times per minute. Builds a stretchy nest with plant material and spiderwebs, gluing pieces of lichen to the exterior for camouflage. Attracted to colorful, tubular flowers. Will extract and eat insects trapped in spiderwebs. A long-distance migrator, wintering as far south as the tropics of Central America.

in flight

SUMMER

Green Heron
Butorides virescens

Size: 16–22" (41–56 cm)

Male: Short and stocky. Blue-green back and rust-red neck and breast. Dark-green crest. Short legs are normally yellow but turn bright orange during the breeding season.

Female: same as male

Juvenile: similar to adults, with a bluish-gray back and white-streaked breast and neck

Nest: platform; female and male build; 2 broods per year

Eggs: 2–4; light green without markings

Incubation: 21–25 days; female and male incubate

Fledging: 35–36 days; female and male feed the young

Migration: complete, to Florida and Georgia

Food: small fish, aquatic insects, small amphibians

Compare: Great Blue Heron (p. 247) is much larger and has a long neck. Look for a small heron with a dark-green back stalking wetlands.

Stan's Notes: Often gives an explosive, rasping "skyew" call when startled. Holds its head close to its body, which sometimes makes it look like it doesn't have a neck. Waits on the shore or wades stealthily, hunting for small fish, aquatic insects and small amphibians. Places an object, such as an insect, on the water's surface to attract fish to catch. Nests in a tall tree, often a short distance from the water. The nest can be very high up in the tree. Babies give a loud ticking sound, like the ticktock of a clock.

female
p. 181

male

Wood Duck
Aix sponsa

Size: 17–20" (43–51 cm)

Male: Small, highly ornamented dabbling duck. Mostly green head and crest patterned with black and white. Rusty chest and a white belly. Red eyes.

Female: brown duck with a bright-white eye-ring, not-so-obvious crest and blue patch on wings (speculum), often hidden

Juvenile: similar to female

Nest: cavity; female lines an old woodpecker cavity or a nest box in a tree; 1 brood per year

Eggs: 10–15; creamy white without markings

Incubation: 28–36 days; female incubates

Fledging: 56–68 days; female teaches the young to feed

Migration: non-migrator to partial migrator; moves around to find open water

Food: aquatic insects, plants, seeds

Compare: Male Green-winged Teal (p. 165) is not as colorful. Male Northern Shoveler (p. 259) is larger and has a large spoon-shaped bill.

Stan's Notes: A common duck of quiet, shallow backwater ponds. Nearly went extinct around 1900 due to overhunting, but it's doing well now. Nests in a tree cavity or a nest box in a tree. Seen flying in forests or perching on high branches. Female takes off with a loud squealing call and enters the nest cavity from full flight. Lays some eggs in a neighboring nest (egg dumping), resulting in more than 20 eggs in some clutches. Hatchlings stay in the nest for 24 hours, then jump from as high as 60 feet (18 m) to the ground or water to follow their mother. They never return to the nest.

female
p. 185

male

Mallard
Anas platyrhynchos

YEAR-ROUND

Size: 19–21" (48–53 cm)

Male: Large, bulbous green head, white necklace and rust-brown or chestnut chest. Gray-and-white sides. Yellow bill. Orange legs and feet.

Female: brown with an orange-and-black bill and blue-and-white wing mark (speculum)

Juvenile: same as female but with a yellow bill

Nest: ground; female builds; 1 brood per year

Eggs: 7–10; greenish to whitish, unmarked

Incubation: 26–30 days; female incubates

Fledging: 42–52 days; female leads the young to food

Migration: non-migrator to partial migrator

Food: seeds, plants, aquatic insects; will come to ground feeders offering corn

Compare: Most people recognize this common duck. The male Northern Shoveler (p. 259) has a white chest with rust on sides and a dark spoon-shaped bill. Look for the green head and yellow bill to identify the male Mallard.

Stan's Notes: A familiar dabbling duck of lakes and ponds. Also found in rivers, streams and some backyards. Tips forward to feed on vegetation on the bottom of shallow water. The name "Mallard" comes from the Latin word *masculus*, meaning "male," referring to the male's habit of taking no part in raising the young. Male and female have white underwings and white tails, but only the male has black central tail feathers that curl upward. Unlike the female, the male doesn't quack. Returns to its birthplace each year.

female
p. 187

male

Northern Shoveler
Anas clypeata

Size: 19–21" (48–53 cm)

Male: Medium-size duck with an iridescent green head, rust sides, white chest. Extraordinarily large, spoon-shaped bill, almost always held pointed toward the water.

Female: brown and black, with a green wing patch (speculum) and large, spoon-shaped bill

Juvenile: same as female

Nest: ground; female builds; 1 brood per year

Eggs: 9–12; olive without markings

Incubation: 22–25 days; female incubates

Fledging: 30–60 days; female leads the young to food

Migration: complete, to Kentucky, southern states, Mexico and Central America

Food: aquatic insects, plants

Compare: The male Mallard (p. 257) is similar but lacks the large, spoon-shaped bill. The male Wood Duck (p. 255) is smaller and has a crest.

Stan's Notes: One of several species of shovelers. Called "Shoveler" due to the peculiar, shovel-like shape of its bill. Given the common name "Northern" because it is the only species of these ducks in North America. Seen in shallow wetlands, ponds and small lakes in flocks of 5–10 birds. Flocks fly in tight formation. Swims low in water, pointing its large bill toward the water as if it's too heavy to lift. Usually swims in tight circles while feeding. Feeds mainly by filtering tiny aquatic insects and plants from the surface of the water with its bill. Female gathers plant material and forms it into a nest a short distance from the water.

female
p. 291

male

American Redstart
Setophaga ruticilla

SUMMER MIGRATION

Size: 5" (13 cm)

Male: Striking black warbler with orange patches on the sides, wings and tail. White belly.

Female: olive-brown with yellow patches on the sides, wings and tail, white belly

Juvenile: same as female; male attains orange tinges in the second year

Nest: cup; female builds; 1 brood per year

Eggs: 3–5; off-white with brown markings

Incubation: 12 days; female incubates

Fledging: 9 days; female and male feed the young

Migration: complete, to Florida, Mexico, Central America and South America

Food: insects, seeds, occasionally berries

Compare: The male Baltimore Oriole (p. 263) and male Red-winged Blackbird (p. 31) are much larger. The male American Redstart is the only small black-and-orange bird flitting around the top of trees.

Stan's Notes: This is a common and widespread warbler in Kentucky, mostly seen during spring migration. Found in woodlands, parks and yards and at forest edges. Prefers large, unbroken tracts of forest. Appears hyperactive when it feeds, hovering and darting back and forth to glean insects from leaves. Often droops wings and fans tail before launching out to catch an insect. Look for the flashing black-and-orange colors of the male high up in trees. First-year males have yellow markings and look like the females. Sings a high-pitched song that builds in intensity and then suddenly ends.

female
p. 299

male

Baltimore Oriole

Icterus galbula

SUMMER

Size: 7–8" (18–20 cm)

Male: Flaming orange with a black head and back. White-and-orange wing bars. Orange-and-black tail. Gray bill and dark eyes.

Female: pale yellow with orange tones, gray-brown wings, white wing bars, gray bill, dark eyes

Juvenile: same as female

Nest: pendulous; female builds; 1 brood per year

Eggs: 4–5; bluish with brown markings

Incubation: 12–14 days; female incubates

Fledging: 12–14 days; female and male feed the young

Migration: complete, to Mexico, Central America and South America

Food: insects, fruit, nectar; comes to nectar, orange-half and grape-jelly feeders

Compare: The male Orchard Oriole (p. 265) is much darker orange. Male American Redstart (p. 261) is smaller and has more black than orange. Look for the flaming orange to identify the male Baltimore Oriole.

Stan's Notes: A fantastic songster, often heard before seen. Easily attracted to a feeder that offers sugar water (nectar), orange halves or grape jelly. Parents bring their young to feeders. Hunts at the top of trees, feeding on caterpillars. Female builds a sock-like nest at the outermost branches of tall trees. Prefers parks, yards and forests and often returns to the same area year after year. Young males turn orange-and-black at 1½ years of age. Some of the last birds to arrive in spring (April) and first to leave in fall (August).

female
p. 301

male

first-year
male

Orchard Oriole

Icterus spurius

SUMMER

Size: 7–8" (18–20 cm)

Male: Dark orange with black head, throat, upper back, wings and tail. White wing bar. Bill is long and thin. Gray mark on lower bill.

Female: olive-green back, dull-yellow belly and gray wings with 2 indistinct white wing bars

Juvenile: same as female; first-year male looks like the female, with a black bib

Nest: pendulous; female builds; 1 brood per year

Eggs: 3–5; pale blue to white, brown markings

Incubation: 11–12 days; female and male incubate

Fledging: 11–14 days; female and male feed the young

Migration: complete, to Mexico and Central and South America

Food: insects, fruit, nectar; comes to nectar, orange-half and grape-jelly feeders

Compare: The male Baltimore Oriole (p. 263) is brighter orange. Look for the dark-orange plumage to identify the male Orchard Oriole.

Stan's Notes: Named "Orchard" for its preference for orchards. Also likes open woods. Eats insects until wild fruit starts to ripen. Often nests alone; sometimes nests in small colonies. Parents bring their young to bird feeding stations after they fledge. Many people don't see these birds at their feeders very much during the summer and think they have left, but the birds are still there, hunting for insects to feed to their young. One of the last birds to arrive in spring and one of the first to leave in fall. Spends only four to five months in Kentucky. Often migrates in flocks with Baltimore Orioles.

male

female
p. 101

yellow
male

House Finch

Haemorhous mexicanus

YEAR-ROUND

Size: 5" (13 cm)

Male: Small finch with a red-to-orange face, throat, chest and rump. Brown cap. Brown marking behind eyes. White belly with brown streaks. Brown wings with white streaks.

Female: brown with a heavily streaked white chest

Juvenile: similar to female

Nest: cup, sometimes in cavities; female builds; 2 broods per year

Eggs: 4–5; pale blue, lightly marked

Incubation: 12–14 days; female incubates

Fledging: 15–19 days; female and male feed the young

Migration: non-migrator to partial migrator; will move around to find food

Food: seeds, fruit, leaf buds; visits seed feeders and feeders that offer grape jelly

Compare: The male Purple Finch (p. 269) has a red cap. Look for the brown cap and streaked belly to help identify the male House Finch.

Stan's Notes: A relatively new bird to Kentucky (first reported in the early 1970s and first nest reports by 1981). Can be a common bird at your feeders. Very social, visiting feeders in small flocks. Likes to nest in hanging flower baskets. Male sings a loud, cheerful warbling song. It was originally introduced to Long Island, New York, from the western U.S. in the 1940s and is now found throughout the country. Suffers from a disease that causes the eyes to crust, resulting in blindness and death. Rarely, males are yellow (inset), perhaps due to poor diet.

female
p. 115

male

Purple Finch
Haemorhous purpureus

WINTER

Size: 6" (15 cm)

Male: Raspberry-red head, cap, chest, back and rump. Brownish wings and tail. Large bill.

Female: heavily streaked brown-and-white bird with bold white eyebrows

Juvenile: same as female

Nest: cup; female and male build; 1 brood per year

Eggs: 4–5; greenish blue with brown markings

Incubation: 12–13 days; female incubates

Fledging: 13–14 days; female and male feed the young

Migration: irruptive; moves around in search of food

Food: seeds, insects, fruit; comes to seed feeders

Compare: The male House Finch (p. 267) has a brown cap and a streaked belly. Look for the raspberry cap to help identify the male Purple Finch.

Stan's Notes: Usually seen only during the winter in Kentucky, when flocks of Purple Finches leave their homes farther north and move around searching for food. Travels in flocks of up to 50 birds. Visits seed feeders along with House Finches, which makes it hard to tell them apart. Feeds mainly on seeds; ash tree seeds are an important source of food. Found in coniferous forests, mixed woods, woodland edges and suburban backyards. Flies in the typical undulating, up-and-down pattern of finches. Sings a rich, loud song. Gives a distinctive "tic" note only in flight. Male is not purple. The Latin species name *purpureus* means "purple" (or other reddish colors).

female
p. 297

male

Scarlet Tanager
Piranga olivacea

Size: 7" (18 cm)

Male: Bright scarlet with coal-black wings and tail. Ivory bill and dark eyes.

Female: drab greenish yellow with olive wings and tail, whitish wing linings and dark eyes

Juvenile: same as female

Nest: cup; female builds; 1 brood per year

Eggs: 4–5; blue-green with brown markings

Incubation: 13–14 days; female incubates

Fledging: 9–11 days; female and male feed the young

Migration: complete, to Central and South America

Food: insects, fruit

Compare: Male Summer Tanager (p. 273) is slightly larger and rosy red in color, compared to the scarlet red of male Scarlet Tanager. Male Northern Cardinal (p. 275) is larger, with a black mask and red bill.

Stan's Notes: A tropical-looking bird. Found in mature deciduous woodlands, where it hunts for insects high up in trees. Requires a territory covering at least 4 acres (1.5 ha) for nesting but prefers 8 acres (3 ha). Arrives late in spring and leaves early in fall. Male and female both sing like American Robins, but the tanagers intersperse an unusual "chick-burr" call in their songs. The song of the female is like that of the male, only softer. This bird is one of hundreds of tanager species in the world. Nearly all are brightly colored and live in the tropics. The name "Tanager" comes from a South American Tupi Indian word meaning "any small, brightly colored bird." The male sheds (molts) his bright-scarlet plumage in the fall, appearing more like the female during winter.

271

female
p. 303

male

SUMMER

Summer Tanager
Piranga rubra

Size: 8" (20 cm)

Male: Bright rosy-red bird with darker red wings.

Female: overall yellow with slightly darker wings

Juvenile: male has patches of red and green over the entire body, female is same as adult female

Nest: cup; female builds; 1–2 broods per year

Eggs: 3–5; pale blue with dark markings

Incubation: 10–12 days; female incubates

Fledging: 12–15 days; female and male feed young

Migration: complete, to Mexico and Central and South America

Food: insects, fruit

Compare: Male Scarlet Tanager (p. 271) is slightly smaller and is scarlet red with black wings, compared to the rosy red of male Summer Tanager. Male Northern Cardinal (p. 275) is similar in size, but it has a black mask, large crest and red bill.

Stan's Notes: Found throughout Kentucky where woodlands exist, especially in mixed pine and oak forests. Due to clearing of land for agriculture, populations have decreased for over a century and especially most recently. Returning to the state in late April and with young hatching in late May, some pairs have two broods per year. While fruit makes up some of the diet, most of it consists of insects such as bees and wasps. Summer Tanagers unfortunately seem to be parasitized by Brown-headed Cowbirds.

female
p. 139

male

juvenile

Northern Cardinal
Cardinalis cardinalis

YEAR-ROUND

Size: 8–9" (20–23 cm)

Male: Red with a black mask that extends from the face to the throat. Large crest and a large red bill.

Female: buff-brown with a black mask, large reddish bill, and red tinges on the crest and wings

Juvenile: same as female but with a blackish-gray bill

Nest: cup; female builds; 2–3 broods per year

Eggs: 3–4; bluish white with brown markings

Incubation: 12–13 days; female and male incubate

Fledging: 9–10 days; female and male feed the young

Migration: non-migrator

Food: seeds, insects, fruit; comes to seed feeders

Compare: Similar size as male Summer Tanager (p. 273), but the male Tanager is rosy red. Male Scarlet Tanager (p. 271) has black wings and tail. Look for the black mask, large crest and red bill to identify the male Northern Cardinal.

Stan's Notes: Seen in a variety of habitats, including parks. Usually likes thick vegetation. One of the few species in which both males and females sing. Can be heard all year. Listen for its "whata-cheer-cheer-cheer" territorial call in spring. Watch for a male feeding a female during courtship. The male also feeds the young of the first brood while the female builds a second nest. Territorial in spring, fighting its own reflection in a window or other reflective surface. Non-territorial in winter, gathering in small flocks of up to 20 birds. Makes short flights from cover to cover, often landing on the ground. *Cardinalis* denotes importance, as represented by the red priestly garments of Catholic cardinals.

in flight

breeding

juvenile

winter

WINTER

Ring-billed Gull
Larus delawarensis

Size: 18–20" (45–51 cm); up to 4' wingspan

Male: White with gray wings, black wing tips spotted with white, and a white tail, seen in flight (inset). Yellow bill with a black ring near the tip. Yellowish legs and feet. In winter, the back of the head and the nape of the neck are speckled brown.

Female: same as male

Juvenile: white with brown speckles and a brown tip of tail; mostly dark bill

Nest: ground; female and male construct; 1 brood per year

Eggs: 2–4; off-white with brown markings

Incubation: 20–21 days; female and male incubate

Fledging: 20–40 days; female and male feed the young

Migration: complete, to Kentucky, southern states and Mexico

Food: insects, fish; scavenges for food

Compare: A large white gull with a black ring around the bill near the tip.

Stan's Notes: A common gull of garbage dumps and parking lots. It's expanding its range and remaining farther north longer during winter due to successful scavenging in cities. One of the most common gulls in the U.S. A three-year gull with different plumages in each of its first three years. Attains the ring on its bill after the first winter and adult plumage in the third year. Defends a small area around the nest, usually only a few feet. Doesn't nest in Kentucky.

in flight

breeding

juvenile

winter

Herring Gull
Larus argentatus

MIGRATION
WINTER

Size: 23–26" (58–66 cm); up to 5' wingspan

Male: White with slate-gray wings. Black wing tips with tiny white spots. Yellow bill with an orange-red spot near the tip of the lower bill (mandible). Pinkish legs and feet. Winter plumage has gray speckles on head and neck.

Female: same as male

Juvenile: mottled brown to gray, with a black bill

Nest: ground; female and male construct; 1 brood per year

Eggs: 2–3; olive with brown markings

Incubation: 24–28 days; female and male incubate

Fledging: 35–36 days; female and male feed the young

Migration: complete, to Kentucky and southern states

Food: fish, insects, clams, eggs, baby birds

Compare: Ring-billed Gull (p. 277) is smaller and has yellowish legs and feet and a black ring on its bill. Look for the orange-red spot on the bill to help identify the Herring Gull.

Stan's Notes: A common gull of large lakes. An opportunistic bird, scavenging for human food in dumpsters, parking lots and other places with garbage. Takes eggs and young from other bird nests. Often drops clams and other shellfish from heights to break the shells and get to the soft interior. Nests in colonies, returning to the same site annually. Lines its nest with grass and seaweed. It takes about four years for the juveniles to obtain adult plumage. Adults have spotted heads during winter.

in flight

Great Egret
Ardea alba

Size: 36–40" (91–102 cm); up to 4½' wingspan

Male: Tall, thin, all-white bird with a long neck and a long, pointed yellow bill. Black, stilt-like legs and black feet.

Female: same as male

Juvenile: same as adults

Nest: platform; male and female construct; 1 brood per year

Eggs: 2–3; light blue without markings

Incubation: 23–26 days; female and male incubate

Fledging: 43–49 days; female and male feed the young

Migration: complete, to coastal southern states, Mexico and Central America

Food: small fish, aquatic insects, frogs, crayfish

Compare: Smaller in size and similar in shape to the Great Blue Heron (p. 247).

Stan's Notes: Slowly stalks shallow ponds, lakes and wetlands in search of small fish to spear with its long, sharp bill. Gives a loud, dry croak if disturbed or when squabbling for a nest site at the colony. The name "Egret" comes from the French word *aigrette*, meaning "ornamental tufts of plumes." The plumes grow near the tail during the breeding season. Hunted to near extinction in the 1800s and early 1900s for its long plumes, which were used to decorate women's hats. Today, the egret is a protected species.

male

female

SUMMER

Yellow Warbler
Setophaga petechia

Size: 5" (13 cm)

Male: Yellow with thin orange streaks on the chest and belly. Long, pointed dark bill.

Female: same as male but lacks orange streaks

Juvenile: similar to female but much duller

Nest: cup; female builds; 1 brood per year

Eggs: 4–5; white with brown markings

Incubation: 11–12 days; female incubates

Fledging: 10–12 days; female and male feed the young

Migration: complete, to Mexico, Central America and South America

Food: insects

Compare: The Yellow-rumped Warbler (p. 209) has just patches of yellow. The male American Goldfinch (p. 295) has a black forehead and black wings. The female American Goldfinch (p. 295) has white wing bars. Look for the orange streaks on the chest to identify the male Yellow Warbler.

Stan's Notes: A widespread and common warbler in Kentucky. Seen in shrubby areas close to water, gardens and backyards. Zooms around shrubs and shorter trees. A prolific insect eater, gleaning caterpillars and other insects from tree leaves. Male sings loudly, flies off to grab a bug, and then starts singing again. Male sings a string of notes that sound like "sweet, sweet, sweet, I'm-so-sweet!" Males arrive in spring before females to claim territories. Migrates at night in mixed flocks of warblers. Rests and feeds during the day. Starts to migrate south in August, returning to Kentucky in the latter part of April.

Kentucky Warbler
Geothlypis formosa

Size: 5" (13 cm)

Male: Bright-yellow eye "spectacles," chin, chest, belly and extending to underneath the tail. Black crown with black extending down the sides of neck. Dark olive back and wings. A characteristic short tail and long legs.

Female: same as male, but black areas not as dark

Juvenile: similar to female

Nest: cup; female and male construct; 1–2 broods per year

Eggs: 3–6; white with dark markings

Incubation: 12–13 days; female incubates

Fledging: 8–10 days; female and male feed young

Migration: complete, to the Caribbean, Central America and South America

Food: insects

Compare: Magnolia Warbler (p. 287) has a gray crown and white eyebrows, unlike the black cap and yellow "spectacles" of Kentucky Warbler.

Stan's Notes: One of the most common forest-dwelling birds in Kentucky, but can be very shy and usually heard more than seen. Feeds mostly on the ground, hopping along and chasing after any insects. Builds cup nest of dead leaves, grasses and rootlets under logs or in shrubs. Can be very sensitive to deforestation. A frequent Brown-headed Cowbird host. The Kentucky Warbler was named by ornithologist Alexander Wilson in 1810 and named after the area where he found it.

male

winter male

female

MIGRATION

Magnolia Warbler
Setophaga magnolia

Size: 5" (13 cm)

Male: Yellow and black with a gray crown and white eyebrows. Heavy black streaks on a yellow chest and belly. White wing patch. Yellow rump. Obvious white patches on tail.

Female: similar to male but lacks black on the face and has 2 white wing bars

Juvenile: same as female

Nest: cup; female and male build; 1 brood per year

Eggs: 3–5; white with brown markings

Incubation: 11–13 days; female incubates

Fledging: 8–10 days; female and male feed the young

Migration: complete, to Mexico and Central America

Food: insects

Compare: Yellow-rumped Warbler (p. 209) has less yellow. Kentucky Warbler (p. 285) has a black crown and yellow eye "spectacles." Prairie Warbler (p. 289) has an olive back with chestnut streaks. The Yellow-throated Warbler (p. 211) has a white belly with black streaks.

Stan's Notes: Common in the state during spring (mid-May) and fall (late August) migrations. Populations are stable due to the ability to adapt to second-growth forest. Look for it low in trees, where it feeds on insects. Often fans its tail while picking insects from the undersides of leaves. Males often feed higher up in trees than the females. Named "Magnolia" when ornithologist Alexander Wilson saw the species for the first time in a magnolia tree.

Prairie Warbler
Setophaga discolor

Size: 5" (13 cm)

Male: Olive back with chestnut-colored streaks. Bright yellow from the chin to belly. Black streaks on sides from neck down. Black line through eyes. Yellow eyebrows.

Female: same as male, only duller

Juvenile: similar to female

Nest: cup; female builds; 2 broods per year

Eggs: 3–5; white with brown markings

Incubation: 11–14 days; female incubates

Fledging: 8–11 days; female and male feed young

Migration: complete, to the Caribbean

Food: insects

Compare: The Magnolia Warbler (p. 287) has more black, and the Yellow Warbler (p. 283) has none. Common Yellowthroat (p. 293) has a complete black mask. Watch for Prairie Warbler to twitch its tail when feeding.

Stan's Notes: A common and widespread warbler in Kentucky. Returns to the state in mixed flocks of warblers in late April to mid-May. It was unfortunately misnamed "Prairie" when first found in a barren area in Kentucky. Nests in dry, brushy clearings and forest edges, making it a perfect host for Brown-headed Cowbirds. Will sometimes desert a parasitized nest. Nests in upright fork of a tree. Feeds young mainly caterpillars.

male
p. 261

female

American Redstart
Setophaga ruticilla

SUMMER
MIGRATION

Size: 5" (13 cm)

Female: Olive-brown warbler with yellow patches on the sides, wings and tail. White belly.

Male: black with orange patches on the sides, wings and tail; white belly

Juvenile: same as female; male attains orange tinges in the second year

Nest: cup; female builds; 1 brood per year

Eggs: 3–5; off-white with brown markings

Incubation: 12 days; female incubates

Fledging: 9 days; female and male feed the young

Migration: complete, to Florida, Mexico, Central America and South America

Food: insects, seeds, occasionally berries

Compare: Female Yellow-rumped Warbler (p. 209) has a yellow patch on its rump. Look for yellow patches on the sides, wings and tail to help identify the female Redstart.

Stan's Notes: This is a common and widespread breeding warbler in Kentucky. Found in woodlands, parks and yards and at forest edges. Prefers large, unbroken tracts of forest. Appears hyperactive when it feeds, hovering and darting back and forth to glean insects from leaves. Often droops wings and fans tail before launching out to catch an insect. Look for the flashing black-and-orange colors of the male high up in trees. First-year males have yellow markings and look like the females. Sings a high-pitched song that builds in intensity and then suddenly ends.

Common Yellowthroat
Geothlypis trichas

Size: 5" (13 cm)

Male: Olive-brown with a bright-yellow throat and chest, a white belly and a distinctive black mask outlined in white. Long, thin, pointed black bill.

Female: similar to male but lacks a black mask

Juvenile: same as female

Nest: cup; female builds; 2 broods per year

Eggs: 3–5; white with brown markings

Incubation: 11–12 days; female incubates

Fledging: 10–11 days; female and male feed the young

Migration: complete, to southern coastal states, Mexico and Central America

Food: insects

Compare: The male American Goldfinch (p. 295) has a black forehead and wings. The male Yellow Warbler (p. 283) has fine orange streaks on chest and lacks the black mask. The Yellow-rumped Warbler (p. 209) only has patches of yellow and lacks the yellow chest of the Yellowthroat.

Stan's Notes: A common warbler of open fields and marshes. Sings a cheerful, well-known "witchity-witchity-witchity-witchity" song from deep within tall grasses. Male sings from prominent perches and while he hunts. He performs a curious courtship display, bouncing in and out of tall grass while singing a mating song. Female builds a nest low to the ground. Young remain dependent on their parents longer than most other warblers. A frequent cowbird host.

American Goldfinch
Spinus tristis

Size: 5" (13 cm)

Male: Canary-yellow finch with a black forehead and tail. Black wings with white wing bars. White rump. No markings on the chest. Winter male is similar to the female.

Female: dull olive-yellow plumage with brown wings; lacks a black forehead

Juvenile: same as female

Nest: cup; female builds; 1 brood per year

Eggs: 4–6; pale blue without markings

Incubation: 10–12 days; female incubates

Fledging: 11–17 days; female and male feed the young

Migration: partial to non-migrator; small flocks of up to 20 birds move around to find food

Food: seeds, insects; will come to seed feeders

Compare: Male Yellow Warbler (p. 283) is all yellow with orange streaking on chest. The Pine Siskin (p. 99) has a streaked chest and belly and yellow wing bars. The female House Finch (p. 101) and female Purple Finch (p. 115) have heavily streaked chests.

Stan's Notes: A common year-round backyard resident. Most often found in open fields, scrubby areas and woodlands. Enjoys Nyjer seed in feeders. Breeds in late summer. Lines its nest with the silky down from wild thistle. Almost always in small flocks. Twitters while it flies. Flight is roller coaster-like. Moves around to find adequate food during winter. Often called Wild Canary due to the male's canary-colored plumage. Male sings a pleasant, high-pitched song. Moves only far enough south to find food.

female

male
p. 271

Scarlet Tanager
Piranga olivacea

Size: 7" (18 cm)

Female: Drab greenish yellow with olive wings and tail. Whitish wing linings. Dark eyes.

Male: bright scarlet with coal-black wings and tail, an ivory bill and dark eyes

Juvenile: same as female

Nest: cup; female builds; 1 brood per year

Eggs: 4–5; blue-green with brown markings

Incubation: 13–14 days; female incubates

Fledging: 9–11 days; female and male feed the young

Migration: complete, to Central and South America

Food: insects, fruit

Compare: Female Summer Tanager (p. 303) is slightly larger, and it is overall mustard-yellow with a larger, thicker bill. The female Baltimore Oriole (p. 299) has gray-brown wings. The female American Goldfinch (p. 295) has white wing bars. Look for the olive wings to identify the female Tanager.

Stan's Notes: A tropical-looking bird found in mature deciduous woodlands, where it hunts for insects high up in trees. Requires a territory covering at least 4 acres (1.5 ha) for nesting but prefers 8 acres (3 ha). Arrives late in spring and leaves early in fall. Both the female and male sing like American Robins, but the tanagers intersperse an unusual "chick-burr" call in their songs. The song of the female is like that of the male, only softer. Nearly all are brightly colored and live in the tropics. The name "Tanager" comes from a South American Tupi Indian word meaning "any small, brightly colored bird."

male
p. 263

female

Baltimore Oriole
Icterus galbula

Size: 7–8" (18–20 cm)

Female: Pale yellow with orange tones and gray-brown wings with white wing bars. Gray bill. Dark eyes.

Male: flaming orange with a black head and back, white-and-orange wing bars, an orange-and-black tail, a gray bill and dark eyes

Juvenile: same as female

Nest: pendulous; female builds; 1 brood per year

Eggs: 4–5; bluish with brown markings

Incubation: 12–14 days; female incubates

Fledging: 12–14 days; female and male feed the young

Migration: complete, to Mexico, Central America and South America

Food: insects, fruit, nectar; comes to nectar, orange-half and grape-jelly feeders

Compare: The female Orchard Oriole (p. 301) has a dull-yellow belly. Often confused with the female Scarlet Tanager (p. 297). Look for the gray-brown wings to identify the female Baltimore Oriole.

Stan's Notes: A fantastic songster, often heard before seen. Easily attracted to bird feeders that offer sugar water (nectar), orange halves or grape jelly. Parents bring young to feeders. Hunts at the top of trees, feeding on caterpillars. Female builds a sock-like nest at the outermost branches of tall trees. Prefers parks, yards and forests and often returns to the same area year after year. Some of the last birds to arrive in spring (April) and first to leave in fall (August).

male
p. 265

female

first-year
male

Orchard Oriole

Icterus spurius

Size: 7–8" (18–20 cm)

Female: Olive-green with a dull-yellow belly. Gray wings with 2 indistinct white wing bars. Long, thin bill with a gray mark on the lower bill.

Male: dark orange with black head, throat, upper back, wings and tail; 1 white wing bar

Juvenile: same as female; first-year male looks like the female, with a black bib

Nest: pendulous; female builds; 1 brood per year

Eggs: 3–5; pale blue to white, brown markings

Incubation: 11–12 days; female and male incubate

Fledging: 11–14 days; female and male feed the young

Migration: complete, to Mexico and Central and South America

Food: insects, fruit, nectar; comes to nectar, orange-half and grape-jelly feeders

Compare: Female Baltimore Oriole (p. 299) is similar, but it has orange tones and more-distinct wing bars. The female Summer Tanager (p. 303) is mustard-yellow with a larger bill. Female Scarlet Tanager (p. 297) has olive wings.

Stan's Notes: Named "Orchard" for its preference for orchards. Also likes open woods. Eats insects until wild fruit starts to ripen. Often nests alone; sometimes nests in small colonies. Parents bring their young to bird feeding stations after they fledge. Many people don't see these birds at feeders much during the summer and think they have left, but the birds are still there, hunting for insects to feed to their young. One of the last birds to arrive in spring and one of the first to leave in fall.

male
p. 273

female

Summer Tanager
Piranga rubra

Size: 8" (20 cm)

Female: Some show a faint wash of red, but most females are a mustard-yellow overall with slightly darker wings.

Male: bright rosy-red bird with darker red wings

Juvenile: male has patches of red and green over the entire body, female is same as adult female

Nest: cup; female builds; 1–2 broods per year

Eggs: 3–5; pale blue with dark markings

Incubation: 10–12 days; female incubates

Fledging: 12–15 days; female and male feed young

Migration: complete, to Mexico and Central and South America

Food: insects, fruit

Compare: Female Scarlet Tanager (p. 297) is slightly smaller, and it has darker wings and a drab greenish-yellow head, back and rump. Female Orchard Oriole (p. 301) and Baltimore Oriole (p. 299) are similar, but they have wing bars. Look for Summer Tanager's lack of wing bars and larger, thicker bill to identify.

Stan's Notes: Found throughout Kentucky where woodlands exist, especially in mixed pine and oak forests. Due to clearing of land for agriculture, populations have decreased for over a century and especially most recently. Returning to the state in late April and with young hatching in late May, some pairs have two broods per year. While fruit makes up some of the diet, most of it consists of insects such as bees and wasps. Summer Tanagers unfortunately seem to be parasitized by Brown-headed Cowbirds.

Eastern Meadowlark
Sturnella magna

Size: 9" (23 cm)

Male: Robin-shaped bird with a brown back and yellow chest and belly. V-shaped black necklace. Short tail with white outer tail feathers, best seen when flying away.

Female: same as male

Juvenile: same as adult

Nest: cup, on the ground in dense cover; female builds; 2 broods per year

Eggs: 3–5; white with brown markings

Incubation: 13–15 days; female incubates

Fledging: 11–12 days; female and male feed the young

Migration: non-migrator; moves around to find food

Food: insects, seeds

Compare: This is the only large yellow bird that has a black V mark on the breast.

Stan's Notes: A songbird of open grassy country, singing when perched and in flight. Given the name "Meadowlark" because it's a bird of meadows and sings like the larks of Europe. Best known for its wonderful, clear, flute-like whistling song. Often seen perching on fence posts but will quickly dive into tall grass when approached. Sometimes domes its nest with dried grass. Not in the lark family. A member of the blackbird family, related to grackles and orioles. Resident birds are joined in winter by migrating birds from the north, swelling populations.

BIRDING ON THE INTERNET

Birding online is a great way to discover additional information and learn more about birds. These websites will assist you in your pursuit of birds. Web addresses sometimes change a bit, so if one no longer works, just enter the name of the group into a search engine to track down the new address.

Site	Address
Author Stan Tekiela's homepage	naturesmart.com
American Birding Association	aba.org
Audubon Society of Kentucky	audubonsocietyofky.org
Central Kentucky Audubon Society	centralkentuckyaudubon.org
The Cornell Lab of Ornithology	birds.cornell.edu
eBird	ebird.org
Kentucky Audubon Council	www.kentuckyauduboncouncil.org
Kentucky Ornithological Society	www.birdky.org
Louisville Audubon Society	www.louisvilleaudubon.org
Raptor Rehabilitation of Kentucky, Inc.	www.raptorrehab.org

CHECKLIST/INDEX BY SPECIES

Use the boxes to check the birds you've seen.

MORE FOR KENTUCKY BY STAN TEKIELA

Identification Guides
Birds of Prey of the Midwest (quick guide)

Birds of Prey of the Midwest Field Guide

Birds of the Midwest

Midwest Birding Companion

Stan Tekiela's Birding for Beginners: Midwest

Wildflowers of the Midwest

Children's Books: Adventure Board Book Series
Floppers & Loppers

Paws & Claws

Peepers & Peekers

Snouts & Sniffers

Children's Books
C is for Cardinal

Can You Count the Critters?

Critter Litter

Children's Books: Wildlife Picture Books
Baby Bear Discovers the World

The Cutest Critter

Do Beavers Need Blankets?

Hidden Critters

Jump, Little Wood Ducks

Some Babies Are Wild

Super Animal Powers

What Eats That?

Whose Baby Butt?

Whose Butt?

Whose House Is That?

Whose Track Is That?

Favorite Wildlife Series
Bald Eagles
Hummingbirds
Loons

Nature Books
Bird Trivia
Start Mushrooming
A Year in Nature with Stan Tekiela

Wildlife Appreciation Series
Backyard Birds
Bears
Bird Migration
Cranes, Herons & Egrets
Deer, Elk & Moose
Wild Birds

Nature Appreciation Series
Bird Nests
Feathers
Wildflowers

Our Love of Wildlife Series
Our Love of Loons
Our Love of Owls

Nature's Wild Cards (playing cards)
Bears
Birds of the Midwest
Hummingbirds
Loons
Mammals of the Midwest
Owls
Raptors
Trees of the Midwest
Wildflowers of the Midwest

ABOUT THE AUTHOR

Naturalist, wildlife photographer and writer Stan Tekiela is the originator of the popular state-specific field guide series that includes the *Birds of Tennessee Field Guide*. Stan has authored more than 190 educational books, including field guides, quick guides, nature books, children's books and more, presenting many species of animals and plants.

With a Bachelor of Science degree in natural history from the University of Minnesota and as an active professional naturalist for more than 30 years, Stan studies and photographs wildlife throughout the United States and Canada. He has received national and regional awards for his books and photographs and is also a well-known columnist and radio personality. His syndicated column appears in more than 25 newspapers, and his wildlife programs are broadcast on a number of Midwest radio stations. You can follow Stan on Facebook and Twitter or contact him via his website, naturesmart.com.